D0549327

**Graham Booth**

Physics

# Contents

## 1 Electric circuits

## 2 Force and motion

## 3 Waves

## 4 The Earth and beyond

## 5 Energy

# Radioactivity

# Electronics

# Using waves

# Forces and their effects

# Particles

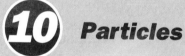

# This book and your GCSE course

| Awarding Body | AQA | EDEXCEL A | EDEXCEL B |
|---|---|---|---|
| Web address | www.aqa.org.uk | www.edexcel.org.uk | |
| Syllabus number | 3451 | 1540 | 1549 |
| Modular tests | None | None | 6 tests each of 30 mins 30% |
| Terminal papers | 1 paper 2½ hours 80% | 1 paper core 1½ hours. 1 paper extension 1 hour 80% | 2 papers core, 30 minutes each. 1 paper extension, 30 minutes 50% |
| Coursework | 20% | 20% | 20% |
| **Core Physics** | | | |
| Electricity | 10.1–10.5 | P1 | 5 and 12 |
| Forces and motion | 10.7–10.9 | P2 | 11 and 12 |
| Waves | 10.13–10.14 and 10.16–10.18 | P3 | 6 and 12 |
| The Earth and beyond | 10.19–10.20 | P4 | 6 and 11 |
| Energy resources and energy transfer | 10.21–10.26 | P1 and P5 | 5 and 11 |
| Radioactivity | 10.27–10.28 | P6 | 6 and 11 |
| **Extension Physics** | | | |
| Electronics | | | |
| Electronic control | 10.6 | | |
| Combining resistors | | | |
| Using waves | | | |
| Refraction and lenses | 10.15 | | |
| Resonance | | | |
| Wave interference | | | |
| Communicating with waves | | P7 | 17 |
| Colour | | | |
| Forces and their effects | | | |
| Projectiles and momentum | 10.11 | | |
| Turning in a circle | 10.10–10.12 | P7 | 17 |
| Some effects of forces | | | |
| Particles | | | |
| Atoms and nuclei | | P8 | 18 |
| Electron beams | | P8 | 18 |
| Particles in motion | | P8 | 18 |

*Visit your awarding body for full details of your course or download your complete GCSE specifications.*

# STAY YOUR COURSE!

Use these pages to get to know your course
- Make sure you know your exam board
- Check which specification you are doing

- Know how your course is assessed:
  – what format are the papers?
  – how is coursework assessed?
  – how many papers?

| OCR A option A | OCR A option B | WJEC | NICCEA |
|---|---|---|---|
| www.ocr.org.uk | | www.wjec.co.uk | www.ccea.org.uk |
| 1982 | | 200 | |
| None | | None | None |
| 1 paper core, 1½ hours. 1 paper extension, 45 minutes 80% | | 1 paper, either 2 hours (foundation) or 2½ hours (higher) 80% | 2 papers, either 1¼ hours (foundation) or 1¾ hours (higher), 75% |
| 20% | | 20% | 25% |
| 4.1 and 4.7 | 4.1 and 4.7 | 1 | 4 |
| 4.2 | 4.2 | 2 | 2 |
| 4.3 and 4.4 | 4.3 and 4.4 | 3 | 3 |
| 4.6 | 4.6 | 4 | 6 |
| 4.2 and 4.7 | 4.2 and 4.7 | 1 and 5 | 1 and 4 |
| 4.5 | 4.5 | 6 | 5 |
| A1 | B3 | 1 | |
| | | 1 | 4 |
| A2 | | | 3 |
| A2 | B3 | | |
| A2 | | 3 | |
| | B2 | | |
| A3 | B1 | 2 | 2 |
| | | 2 | 2 |
| | B1 | 2 | 2 |
| | | 6 | 5 |
| | B2 | 6 | |
| A3 | B1 and B3 | 5 | |

# Preparing for the examination

## Planning your study

The final three months before taking your GCSE examination are very important in achieving your best grade. However, your success can be assisted by an organised approach throughout the course.

- After completing a topic in school or college, go through the topic again in Letts GCSE Physics Guide. Copy out the main points onto a sheet of paper or use a highlighter pen to emphasise them.
- A couple of days later try to write out these key points from memory. Check differences between what you wrote originally and what you wrote later.
- If you have written your notes on a piece of paper, keep this for revision later.
- Try some questions from the book and check your answers.
- Decide whether you have mastered the topic fully and make a note of any weaknesses you think you have. Concentrate on rectifying these weaknesses using your Letts GCSE Physics Guide.

## Preparing a revision programme

At least three months before the final examination go through the list of topics in your Awarding Body's specification. Identify which topics you feel you need to concentrate on. It is a temptation at this time to spend valuable revision time on the things you already know and can do. It makes you feel good but does not move you forward.

When you feel you have mastered all the topics spend time answering practice questions. Each time check your answers with the answers given. In the final couple of weeks go back to your summary sheets (or highlighting in the book).

## How this book will help you

Letts GCSE Physics Guide will help you because:

- it contains the essential content for your GCSE course without the extra material that will not be examined
- it contains Progress checks and GCSE questions to help you to confirm your understanding
- it gives sample GCSE questions with answers and advice from an examiner on how to improve
- examination questions from 2003 will be different from those in 2002 or 2001. Trying past questions may not help you when answering some parts of the questions in 2003. The questions in this book have been written by experienced examiners who are writing the questions for 2003 and beyond
- the summary table will give you a quick reference to the requirements for your examination
- marginal comments and highlighted key points will draw to your attention important things you might otherwise miss.

# Five ways to improve your grade

## 1. Read the question carefully

Many students fail to answer the actual question set. Perhaps they misread the question or answer a similar question they have seen before. Read the question once all the way through and then again more slowly. Some students underline or highlight key words in the question as they read it through. Questions at GCSE contain a lot of information. You should be concerned if you are not using the information in your answer.

## 2. Give enough detail

If a part of a question is worth three marks you should make at least three separate points. Be careful that you do not make the same point three times. Approximately 25% of the marks on your final examination papers are awarded for questions requiring longer answers.

## 3. Quality of Written Communication (QWC)

From 2003 some marks on GCSE papers are given for the quality of your written communication. This includes correct sentence structures, correct sequencing of events and use of scientific words.
Read your answer through slowly before moving on to the next part.

## 4. Correct use of scientific language

There is important Scientific vocabulary you should use. Try to use the correct scientific terms in your answers and spell them correctly. The way in which language is used is often a difference between successful and unsuccessful candidates. As you revise make a list of physical terms you meet and check that you understand the meanings of these words.

## 5. Show your working

All Physics papers include calculations. Many of these rely on recall of physical relationships. You should always state the relationship and show your working in full. Then if you make an arithmetical mistake, you may still receive marks for correct science. However, you will not be awarded any marks if you cannot recall the correct relationship. Check that your answer is given to the correct number of significant figures and give the correct unit.

# Core material

| Topic | Section | Studied in class | Revised | Practice questions |
|---|---|---|---|---|
| 1.1 Current, voltage and resistance | Measuring current and voltage | | | |
| | Current and resistance | | | |
| 1.2 Using mains electricity | Alternating and direct current | | | |
| 1.3 Electric charge | Creating static charge | | | |
| 2.1 Speed, velocity and acceleration | Distance time graphs | | | |
| | Speed, displacement and velocity | | | |
| | Acceleration and graphs | | | |
| 2.2 Movement and force | Starting and stopping | | | |
| | Force and acceleration | | | |
| 2.3 The effects of forces | Forces and materials | | | |
| 3.1 Wave properties and sound | What is a wave? | | | |
| | Wave properties | | | |
| 3.2 Light and the electro-magnetic spectrum | Images from light | | | |
| | A spectrum of waves | | | |
| | Communicating with electromagnetic waves | | | |
| 3.3 The restless Earth | Evidence for the Earth's structure | | | |
| | Movement in the Earth's crust | | | |
| 4.1 The Solar System and its place in the Universe | The Solar System | | | |
| 4.2 Evolution | The life of a star | | | |
| | The past and future of the universe | | | |
| 5.1 Energy transfer and insulation | The nature of the surface | | | |
| | Insulating buildings and bodies | | | |
| 5.2 Work, efficiency and power | Work and energy transfer | | | |
| 5.3 Generating and distributing electricity | The motor effect | | | |
| | Electromagnetic induction | | | |
| | Energy resources | | | |
| 6.1 Ionising radiation | Radiation from the nucleus | | | |
| | The effect on the nucleus | | | |
| 6.2 Using radiation | Radioactive decay and half-life | | | |
| | Nuclear power | | | |
| | Some other uses of radioactivity | | | |

*The following topics are covered in this section:*

- **Current, voltage and resistance**
- **Using mains electricity**
- **Electric charge**

## What you should know already

Use words from the list to complete the passage and label the symbols in the circuit diagram.

You can use each word more than once.

| ammeter | break | cell | circuit | complete | conductors | current | decrease |
|---------|-------|------|---------|----------|------------|---------|----------|
| insulators | lamp | metals | negative | parallel | series | voltmeter | |

A complete current path from the positive to the 1._____ terminals of a battery is called a 2._____. In a circuit, electric current passes in the wires and other components. Materials that allow current to pass in them are called 3._____, those that do not allow current to pass are called 4._____. All 5._____ conduct electricity.

A circuit that has only one path for the current is called a 6._____ circuit, where there is more than one current path the circuit is a 7._____ one. The voltage from a battery or power supply is measured with a 8._____ placed in 9._____. The voltage from a battery can be increased by adding an extra 10._____ in series.

Increasing the voltage of the battery causes the current in a series circuit to increase, but adding extra components such as lamps causes the current to 11._____. The current in a series circuit is measured with an 12._____ placed in 13._____ with the other components. Lamps and other components do not use up 14._____, so it does not matter where an ammeter is placed in a series circuit, as the current has the same value at all points.

Switches have two terminals. In the "on" position they are joined by a conductor, but in the "off" position there is a 15._____ in the circuit so it is not 16._____.

The diagram shows a series circuit.

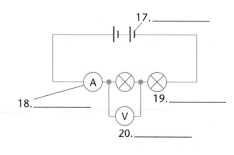

17._____

18._____

19._____

20._____

# 1.1 Current, voltage and resistance

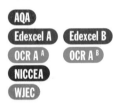

*After studying this section you should be able to:*

● *describe how the current in common circuit components varies with the applied voltage*
● *recall and use the relationship between resistance, current and voltage*
● *explain how electric current is due to charge flow.*

## Measuring current and voltage

AQA
Edexcel A    Edexcel B
OCR A ᴬ    OCR A ᴮ
NICCEA
WJEC

**Current** in a circuit is due to a flow of **charge**. Movement of charge is caused by the force from a **voltage** supply such as a battery or mains supply acting on charged particles that are free to move.

Electric current:

● is measured in **amps** using an **ammeter** placed in **series** with the current being measured
● is not used up by the components in a circuit
● transfers energy from the voltage source to the components.

> The term potential difference is an alternative for voltage.

Voltage:

● is measured in **volts** using a **voltmeter** placed in **parallel** with the voltage source or component
● is a measure of the **energy transfer** from a source or to a component.

### Series and parallel

The diagram, **Fig. 1.1**, shows the symbols used to represent some circuit components.

> The symbols for a diode and a light-dependent resistor sometimes have circles around them – either symbol is acceptable.

| cell | battery | ammeter | voltmeter | fixed resistor | variable resistor |

| fuse | switch | diode | light-dependent resistor | thermistor |

**Fig. 1.1**

A **series circuit** has only one **current path**.

The diagram, **Fig. 1.2**, shows two lamps connected in **series** to a battery, together with an ammeter that measures the current in the circuit and a voltmeter to measure the voltage across one of the lamps.

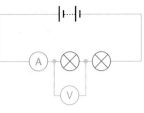

**Fig. 1.2**

In a series circuit:

- the **current** is the **same** at all points in the current path
- the sum of the voltages across the individual components is equal to the voltage of the power supply.

A **parallel circuit** has more than one **current path**. The diagram, **Fig. 1.3**, shows two lamps connected in **parallel** to a battery, together with an ammeter that measures the total current in both lamps.

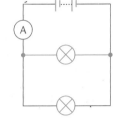

**Fig. 1.3**

In a parallel circuit:

- all components in **parallel** have the **same voltage** across them
- the current splits and rejoins at the junctions
- the total current passing into a junction is equal to the current passing out of the junction.

## Current and resistance

The **current** in a circuit or a component depends on the voltage and also the **resistance**. Resistance is a measure of the opposition to electric current. The higher the resistance of a component, the less current that passes in it for a given voltage.

> **KEY POINT**
> The relationship between voltage, *V*, current, *I*, and resistance, *R* is:
> voltage = current × resistance
> In symbol form this can be written as $V = I \times R$ or $I = V/R$ or $R = V/I$.
> The unit of resistance is the ohm ($\Omega$).

When components are connected in series the **total** resistance is equal to the **sum** of the resistances of the individual components.

### Circuit components

Common circuit components include resistors, lamps and diodes:

- the resistance of a **resistor** such as a **metal wire** does not change provided that there is no significant change in its temperature; a graph of **current** against **voltage** shows that the current is proportional to the voltage
- the wire in a **filament lamp** becomes hotter as the current in the filament increases, causing an increase in its resistance
- a **diode** only allows current to pass in one direction (shown by the direction of the arrow on its symbol).

The diagram, **Fig. 1.4**, shows the variation of current with applied voltage for these components.

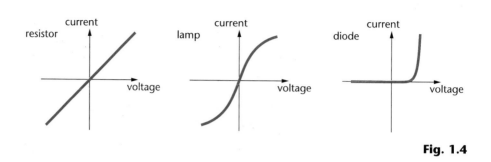

Fig. 1.4

The resistance of some circuit components depends on their surroundings; these components are often found in electronic circuits used for switching and maintaining constant environmental conditions in, for example, greenhouses and incubators:

● the resistance of a **light-dependent resistor (LDR)** decreases with increasing light level
● the resistance of a **thermistor** decreases with increasing temperature.

The graphs in **Fig. 1.5** show the variation of resistance with environmental conditions for these components.

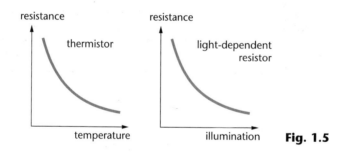

Fig. 1.5

## A current model

Free electrons give metals the properties of being good electrical and thermal conductors.

One difference between a metallic solid such as copper and a non-metallic solid such as salt is that the metal contains charged particles that are moving within the body of the metal. The structure of a metal is one of fixed positive ions surrounded by a "sea" of negatively charged **free electrons**. The free electrons are said to be "free" because they have enough energy to escape from the attraction of the nucleus and move with a random motion, changing speed and direction whenever they collide with a metal ion.

Applying a voltage to the metal causes the free electrons to "drift" slowly in the direction from negative to positive, so there is an overall movement of **charge** in this direction. It is this flow of charge that forms the electric current.

> **KEY POINT**
> The relationship between charge flow and current is:
> charge = current × time
> $Q = I \times t$
> where the charge, $Q$, is measured in coulombs (C) when the current is in amps and the time is in seconds.

**Fluorescent lamps and street lights are examples of gases conducting electricity.**

In a conducting gas or molten or dissolved electrolyte the charge flow is due to the movement of both positively and negatively charged particles, moving in opposite directions.

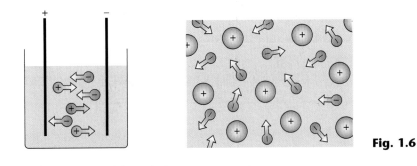

**Fig. 1.6**

## Understanding voltage

**Provided that the connecting wires have a very low resistance, very little energy is transferred to them and they stay cool.**

As charged particles move around a circuit, they gain energy from the battery or power supply and lose it to the components. The voltage across a power supply represents the energy transferred **to** each coulomb of charge from the power supply, and the voltage across a component represents the energy transfer **from** each coulomb of charge to the component.

>
> **KEY POINT**
>
> The relationship between voltage, *V*, charge, *Q*, and the energy transferred, *E*, to or from the charge is:
> voltage = energy transfer ÷ charge passed
> *V = E/Q*
> This relationship also shows the equivalence of the units:
> **1** volt = **1** joule **per** coulomb

**PROGRESS CHECK**

1. Which circuit component has a resistance that depends on the amount of light falling on it?
2. A current of 2.5 A passes in a wire when the voltage across it is 12.5 V. Calculate the resistance of the wire.
3. Name the charged particles that carry the current in a metal wire.

1 . A light-dependent resistor (LDR); 2. 5.0 Ω; 3. Electrons.

# 1.2 Using mains electricity

LEARNING
SUMMARY

*After studying this section you should be able to:*

● **recall the difference between alternating current and direct current**
● **explain the functions of the wires that connect an appliance to the mains supply**
● **describe how electricity from the mains supply is used and measured.**

## Alternating and direct current

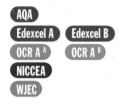

All circuits need a source of energy. Batteries are used in appliances such as torches and portable music players. Batteries use expensive chemicals, so the mains electricity supply is used for appliances used in and around the home. The current in a battery-powered circuit is **direct current (d.c.)**, while that in a mains-powered circuit is **alternating current (a.c.)**:

● direct current passes in the same direction
● alternating current changes direction.

The graphs, **Fig. 1.7**, show the variation of an alternating current and a direct current.

> A direct current does not have to maintain a constant value. The size of the current can change, but the direction is always the same.

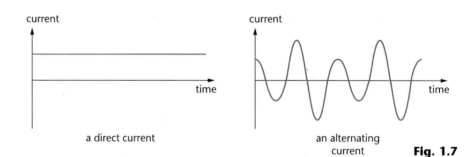

a direct current

an alternating current

**Fig. 1.7**

### Connecting to the mains supply

The flexible cable used to connect an appliance to the mains supply contains either two or three wires within the layers of insulation. All appliances need both a **live** and a **neutral** connection. In addition, those with metal cases or metal parts which could come into contact with the live connection need an **earth** wire.

The diagram, **Fig. 1.8**, shows how a kettle is connected to a mains plug.

> Each wire in the mains cable has two layers of insulation. The colour of the inner layer is different for each conductor. Notice that the cable grip should clamp the outer insulation.

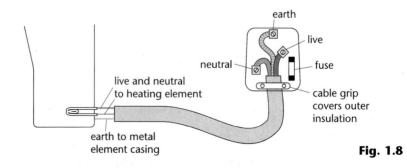

earth

live

neutral

fuse

live and neutral to heating element

cable grip covers outer insulation

earth to metal element casing

**Fig. 1.8**

When the kettle is switched on:

- energy is supplied through the **live** conductor (brown); the voltage of this alternates between positive and negative
- the **neutral** conductor (blue) is there to form a complete current path or circuit
- the **earth** wire (green and yellow stripes) is needed for safety
- the **fuse** is also for safety, it breaks the circuit if the current becomes abnormally high.

The current rating of the fuse should be slightly higher than the current when the appliance is operating normally. A fault such as a **short-circuit** would increase the current passing in the appliance and the wires, causing them to overheat and creating a fire hazard. The fuse prevents this by melting and breaking the circuit when the current exceeds the rating of the fuse.

A fault that causes the metal casing to become live could electrocute the user. The earth wire and fuse together guard against this. If this happens the low resistance path from live to earth causes a large current in the fuse, breaking the circuit. It is important that the fuse is placed in the live conductor so that it cuts off the voltage supply.

> **If the appliance has a plastic casing, the casing cannot conduct electricity.**

Some appliances are said to be **double-insulated** and do not need an earth connection. A double-insulated appliance:

- has a plastic casing with no exposed metal parts
- cannot cause electrocution since the casing cannot become live
- carries this symbol:

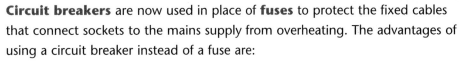

**Circuit breakers** are now used in place of **fuses** to protect the fixed cables that connect sockets to the mains supply from overheating. The advantages of using a circuit breaker instead of a fuse are:

- a circuit breaker is easily reset when a fault has been rectified
- a circuit breaker acts faster than a fuse, giving greater protection.

> **A person stood inside has a much greater resistance than one outside because there is more insulation between them and the ground.**

When using appliances such as lawnmowers and hedge trimmers outdoors, there is an increased risk of electrocution. This is because these appliances can readily cut through a cable and there is little insulation between a person and the earth. A **residual current circuit breaker (RCCB)** should always be used to connect the appliance to the mains supply. An RCCB detects any difference between the current in the live and neutral wires. If the current in the neutral wire is less than that in the live, which happens if current passes in a person, then it cuts off the supply voltage.

## Energy transfer and cost

> **Sound from a television or radio is produced by the movement of a loudspeaker cone.**

Electrical appliances used at home transfer energy from the mains supply to:

- heat
- light
- movement (including sound).

The power of an appliance is the rate at which it transfers energy.

Electrical power is calculated using the relationship:
power = current × voltage
$P = I \times V$
Power is measured in watts (W) where 1 W = 1 J/s.

Appliances used for heating have a much higher power rating than those used for lighting or to reproduce sound. Heating can come about in a number of different ways:

> All wires have some resistance. The greater the resistance, the greater the heating when a given current passes.

- by **convection currents**, when a hot wire heats the surrounding air or water, as in a kettle
- by **infra-red radiation** given off by a wire hot enough to glow red, as in a toaster
- by absorption of **microwaves** by water molecules in food, as in a microwave cooker.

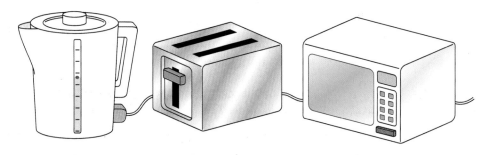

**Fig. 1.9**

The quantity of **energy** transferred from the mains by an appliance depends on:

- the **power** of the appliance

- the **time** for which it is switched on.

> Since $P = IV$ this is also equivalent to $E = IVt$

**KEY POINT**

Energy transfer from electricity is calculated using:
energy = power × time
$E = P \times t$
The energy transfer is in J when the power is measured in W and the time is measured in s.

The electricity meter in your home measures the energy transferred from the mains supply. If it measured the energy in joules it would need a lot of digits as the joule is a very small unit of energy compared to the amount transferred to a typical house each day. Instead of the joule, electricity companies measure the energy supplied in **kilowatt-hours** (kW h).

**KEY POINT**

One kilowatt-hour is the energy supplied to a 1 kW appliance when it operates for 1 hour.

To calculate energy transfer in kW h:

- the equation *energy = power × time* is used

- the power is measured in **kilowatts (kW)**

- the time is measured in **hours (h)**.

The cost of each kilowatt-hour of energy from the electricity mains supply varies but it is currently about 7p. An electricity bill is calculated by multiplying the number of "units" supplied by the cost of each one.

> **The term "unit" is sometimes used to refer to 1 kilowatt-hour.**

> **KEY POINT**
>
> The cost of energy from electricity is calculated using:
> cost = power in kW × time in h × cost of 1 kW h
> or
> cost = number of kW h × cost of 1 kW h

> **PROGRESS CHECK**
>
> 1. Which conductor in a mains lead:
>    (a) supplies the energy?
>    (b) is at a varying voltage?
>    (c) does not normally carry any current?
> 2. The current in a kettle element is 9.5 A when the voltage across it is 240 V. Calculate the power of the kettle element.
> 3. An 8 kW shower is used for a total of 1.5 hours. Calculate the cost of the energy transferred to the shower, if the cost of 1 kW h is 7p.
>
> 1(a) live;  (b) live;  (c) earth;   2. 2280 W;   3. 84p

# 1.3 Electric charge

> **LEARNING SUMMARY**
>
> *After studying this section you should be able to:*
> - *describe how an insulating material can be charged*
> - *explain the forces between charged objects*
> - *recall some everyday uses and hazards of static charge.*

## Creating static charge

You learned in section 1.2 that electric **current** is moving **charge**. A concentration of electric charge that is not moving is called **static charge** or **electrostatic**. A build-up of static charge causes effects that can be hazardous or useful.

Electrons can move freely through a **conductor** but no charged particles are able to move through an **insulator**. An insulator is easily charged by rubbing:

- when an insulator is rubbed with a cloth, the friction forces between the insulator and the cloth cause electrons to be transferred from one to the other

- the object that gains electrons becomes **negatively** charged

- the object that loses electrons becomes **positively** charged.

The diagram, **Fig. 1.10**, shows the imbalance of charge created when a polythene rod is rubbed with a duster. If a conductor such as a metal rod is rubbed while being held in the hand, any unbalanced charge is neutralised by electrons passing through the body between the rod and the earth.

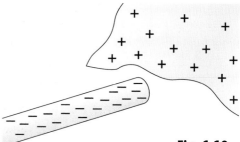

**Fig. 1.10**

> The body acts as a good conductor of small amounts of charge.

A conductor needs to be well-insulated from the earth for it to become charged.

If two balloons are rubbed with a cloth and then placed side-by-side, they push away from each other. Objects with opposite charges, such as one of the charged balloons and the cloth, are pulled towards each other. This shows that:

> The forces between two charged objects are equal in size and act in opposite directions.

> **KEY POINT**
> Objects with similar charges (both negative or both positive) repel each other and those with opposite charges (negative and positive) attract each other.

The existence of repulsive forces between similar charges explains why you experience a shock if you touch an earthed conductor such as the screw on a light switch after walking across a synthetic carpet:

- synthetic carpets, for example those made out of nylon, are good insulators
- charge builds up on the body when walking across the carpet due to the friction forces between shoes and the carpet
- the similar charges repel each other but they cannot leave the body through the carpet which is a very good electrical insulator

> The direction of electron movement depends on whether the body is positively or negatively charged.

- when the body is placed in electrical contact with the earth, electrons move between the body and the earth to discharge the body, creating a current which causes a shock.

## Using static charge

Electrostatic forces are used in photocopiers, inkjet printers and to paint the metal panels used in cars and washing machines.

In a **photocopier**:

- a rubber belt is coated with a material that is a conductor only when illuminated
- the belt is given a positive charge
- an image of the sheet of paper being copied is projected onto the belt, causing the illuminated parts to discharge
- the belt is sprayed with a black powder that is attracted to the charged areas

> The paper needs to be charged to attract the powder from the belt.

- the powder is transferred to a charged sheet of paper which is then heated so that the powder sticks to it.

The diagram, **Fig. 1.11**, shows the discharge of the belt when an image is projected onto it.

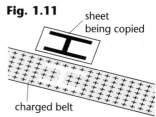

**Fig. 1.11**
sheet being copied
charged belt

In an **inkjet printer**:

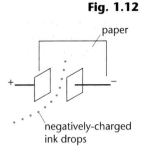

**Fig. 1.12**

paper

- ink drops become charged as they pass through a small hole in a nozzle

- the drops pass between two parallel plates; a voltage applied to the plates deflects the drops

- the deflection of the drops can be increased by increasing the voltage and reversed in direction by reversing the voltage.

negatively-charged ink drops

The deflection of charged ink drops is shown in the diagram, **Fig. 1.12**.

Painting metal panels uses **electrostatic induction**:

- the metal panel is connected to earth

- charged paint powder is sprayed onto the panel

- the positive charges on the paint attract electrons from the earth onto the panel

- the paint powder is attracted to the oppositely-charged panel, ensuring even coverage and little waste

- the panel is then baked in an oven to harden the paint.

**Fig. 1.13**

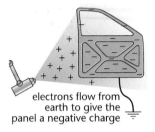

electrons flow from earth to give the panel a negative charge

> If the paint is charged negatively, it repels electrons to earth, leaving the panel positively charged.

Painting by electrostatic induction is shown in the diagram, **Fig. 1.13**.

## Dangers of static charge

Electrostatic charge is dangerous when it causes **lightning** and **sparks** that can ignite fuel.

- when an aircraft is being refuelled with kerosene (paraffin) and when a car is being refuelled with petrol **friction forces** cause **charge separation**

- this could result in a build-up of static charge on the metal frame of the aircraft or metal sleeve of the car refuelling pipe

- if the voltage became high enough to cause a **spark** to earth, it could ignite the fuel

- to prevent this, the framework of an aircraft is connected to **earth** before refuelling and the pipe leading to the petrol tank in a car is connected to the body of the car so that the charge can spread out, preventing the build-up of charge in a small area.

> Charge separation results in the metal frame of an aircraft gaining an opposite charge to the fuel.

**PROGRESS CHECK**

1. Explain why a balloon is attracted to a cloth that it has been rubbed with.
2. Why does a person become charged when walking on a nylon carpet but not when walking on a woollen carpet?
3. How does a metal panel that is connected to earth become charged by electrostatic induction?

1. The cloth and balloon have opposite charges;   2. Nylon is a better insulator than wool;   3. By movement of charge from the person can pass through the woollen carpet to earth;   electrons between the panel and earth.

# Sample GCSE question

1.  The table shows the current that passes in three household appliances when they are connected to the 240 V mains supply.

| Appliance | Current in A |
|---|---|
| grill | 6.0 |
| desk lamp | 0.6 |
| convector heater | 4.5 |

(a) (i) Which of these has the greatest resistance? Explain how you can tell. **[2]**

> *The lamp* ✓ *as it has the least current for the same voltage* ✓*.*

*It is important here to refer to the data in the table and the introduction to the question.*

(ii) Calculate the resistance of the grill. **[3]**

> $R = V/I$ *or resistance = voltage ÷ current* ✓
> $= 240 V ÷ 6.0 A$ ✓
> $= 40 \Omega$ ✓

*These examples show how to set out calculations based on formulas that are either given or that you have to recall. In each case always write down the formula that you are using, then write it with the values in place and finally give the answer with the correct unit.*

(iii) Calculate the power rating of the convector heater. **[3]**

> $P = IV$ *or power = current × voltage* ✓
> $= 4.5 A × 240 V$ ✓
> $= 1080 W$ ✓

(b) The plug fitted to the convector heater has three wires: live, neutral and earth.

(i) Which of these transfers energy from the mains supply? **[1]**

> *The live wire* ✓*.*

(ii) Why is it important that the wires connecting the heater to the mains supply have a low resistance? **[2]**

> *So that the wires do not become hot* ✓*.*
> *Which would waste energy and be a fire hazard* ✓*.*

*The allocation of two marks shows that two separate points are needed. One mark is for realising that the wires would be heated and the second mark is for giving a reason why this is not desirable.*

(iii) Explain how the earth wire, along with the fuse fitted to the plug, guard against electrocution. **[3]**

> *If a fault occurs so that the casing becomes live, a large current passes to earth* ✓*.*
> *This melts the fuse wire* ✓*.*
> *Which breaks the connection in the live wire so that the casing is no longer live* ✓*.*

*This type of question, requiring a full explanation written in a logical order, could carry additional marks for the quality of your written communication. You will not gain these marks if the order of your answer is illogical or your use of sentences is poor.*

# Exam practice questions

**1.** The graph shows how the current in a component varies with the applied voltage.

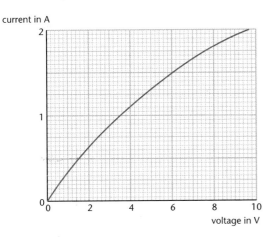

**(a)** What voltage is needed to cause a current of 0.5 A to pass in the component? **[1]**

**(b)** Calculate the resistance of the component when a current of 0.5 A passes in it. **[3]**

**(c)** Suggest what the component could be. Give a reason. **[2]**

**(d)** Explain how the resistance of the component changes as the voltage across it is increased. **[2]**

**2.** When an aircraft is being refuelled, electrostatic charge can build up on the fuel and the airframe.

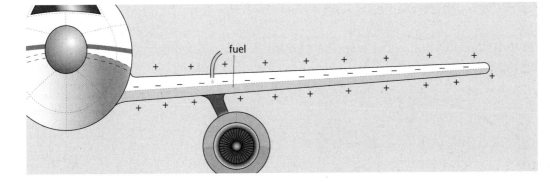

**(a)** **(i)** What charged particle is transferred in this process? **[1]**

**(ii)** What is the sign of the charge on this particle? **[1]**

**(b)** Explain how connecting the airframe to earth prevents the build up of charge. **[3]**

**(c)** Aircraft also become charged as they fly through the air.
Aircraft tyres are made from conducting rubber.
Explain why this is an advantage. **[3]**

# Exam practice questions

**3.** This question is about a car headlamp operating from a 12 V supply.

**(a)** 300 C of charge pass through the filament in one minute.

Calculate the current in the lamp. [3]

**(b)** How much energy is transferred to the filament by this charge passing through it? [3]

**(c)** Calculate the power of the lamp. [3]

**4.** An electric shower has a power of 8.4 kW when it is connected to the 240 V mains supply.

**(a)** Calculate the current in the heater when it is operating normally. [3]

**(b)** Explain why very thick cables are needed to connect it to the mains supply. [2]

**(c)** It is recommended that a residual current circuit breaker (RCCB) is placed in the live supply connection.

What are the advantages of using an RCCB instead of a fuse? [2]

**(d)** The shower is used for 30 minutes each day for the week.

Each kW h of energy costs 7p.

Use the relationship:

$$\text{energy in kW h} = \text{power in kW} \times \text{time in h}$$

to calculate the cost of using the shower each week. [3]

**5.** Here are three graphs that show how the current in a component changes when the voltage is changed.

**(a)** Which graph represents a filament lamp? [1]

**(b)** Which graph represents a fixed resistor at a constant temperature? [1]

**(c)** Which graph represents a component that only allows current to pass in one direction? [1]

**(d)** Which graph represents a diode? [1]

**(e)** The current in a lamp filament is 2.5 A when the voltage across it is 12.0 V.

Calculate the power of the lamp. [3]

# Force and motion

*The following topics are covered in this section:*

- ● *Speed, velocity and acceleration*  ● *Movement and force*
- ● *The effects of forces*

## What you should know already

Use words from the list to complete the passage and label the forces in the diagram.

You can use each word more than once.

| air resistance | anticlockwise | constant | direction | Earth's pull | floating | frictional |
|---|---|---|---|---|---|---|
| gravitational | large | mass | moment | objects | pivot | pressure |
| pulls | resistance | road | time | unbalanced | weight | |

Forces are pushes or 1._____ that are caused by objects and act on other 2._____. Everything on the Earth experiences a downward pull due to the 3._____ attraction between its 4._____ and that of the Earth. The size of this pull is called the object's 5._____.

When an object is not moving, for example a 6._____ ball or a parked car, the forces on it are balanced. Balanced forces also act on objects that move in a straight line at a 7._____ speed. Any change in the motion of an object, such as a change in speed or 8._____, requires an 9._____ force to cause that change.

The speed of a moving object is calculated using the relationship *speed = distance travelled ÷ 10._____ taken*. The units of speed are m/s. Methods of transport that use wheels rely on the 11._____ force between the wheels and 12._____ or track to stop the wheels from slipping and sliding. Parachutes depend on the air 13._____ that acts on all moving objects to limit the maximum speed.

The upward force that acts on a parachutist is due to 14._____ and the downward force is the 15._____.

The turning effect of a force is called its 16._____. The moment depends on the size of the force and the distance to the 17._____; it is calculated using the relationship: *moment = size of force × shortest distance to pivot*. Moments are measured in N m. The principle of moments states that when an object is not turning round, the clockwise and 18._____ moments are balanced.

Pressure describes the effect that a force has in cutting or piercing. The greater the force and the smaller the area, the greater the 19._____. Pressure is calculated using the relationship *pressure = force ÷ area* and measured in units of N/m². Skis have a large area so they exert a small pressure and do not sink in the snow. Knives have a small area so that the 20._____ pressure created can cut through objects.

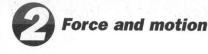
# 2.1 Speed, velocity and acceleration

**LEARNING SUMMARY**

*After studying this section you should be able to:*

● *calculate the speed of an object from a distance–time graph*
● *recall and use the relationship between acceleration, change in velocity and the time taken*
● *interpret velocity–time graphs.*

## Distance–time graphs

AQA
Edexcel A    Edexcel B
OCR A ᴬ    OCR A ᴮ
NICCEA
WJEC

If you go on a journey, the **distance** that you have travelled can only stay the same or increase, so a **distance–time graph** for the journey cannot have a negative slope or **gradient**. The graph, **Fig. 2.1**, shows a distance–time graph for a cycle ride.

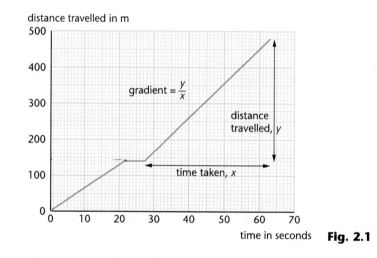

distance travelled in m

gradient = $\frac{y}{x}$

distance travelled, $y$

time taken, $x$

time in seconds    **Fig. 2.1**

> Where the graph line is horizontal, its gradient is zero.

The gradient of the graph line gives information about the cycle ride:

● between 22 s and 27 s the gradient is zero, showing that the cyclist was not moving

● calculating the **speed** of the cyclist using *speed = distance travelled ÷ time taken* is equivalent to calculating the gradient of the graph

● the graph has a steeper gradient between 27 s and 63 s than between 0 s and 22 s, showing a greater speed.

# Speed, displacement and velocity

AQA
Edexcel A   Edexcel B
OCR A ^A    OCR A ^B
NICCEA
WJEC

Knowledge of the speed of an object only tells you how fast it is moving, but its **velocity** also gives information about the **direction** of travel. The direction may be specific, as is the case when describing the velocity of an aircraft, or it may be relative, where velocity in one direction is described as positive and that in the opposite direction is negative.

*Air traffic controllers tell pilots what velocity to fly at. The direction is given in terms of the points of the compass.*

> **KEY POINT**
> It follows that a speed–time graph can only have positive values, but a velocity–time graph can have both positive and negative values, representing motion in opposite directions.

A **displacement–time graph** gives more information than a distance–time graph:

- displacement is the distance an object moves from a fixed position, so it can decrease as well as increase
- displacement can have positive and negative values to show movement in opposite directions
- the gradient of a displacement–time graph represents velocity; a negative gradient represents movement in the opposite direction to that represented by a positive gradient.

# Acceleration and graphs

AQA
Edexcel A   Edexcel B
OCR A ^A    OCR A ^B
NICCEA
WJEC

**Acceleration** involves a **change in velocity**. Speeding up, slowing down and changing direction are all examples of acceleration.

*A negative acceleration in the direction of motion is sometimes called a deceleration; it represents a decrease in speed.*

> **KEY POINT**
> Acceleration is the change in velocity per second. It is calculated using the relationship:
> acceleration = change in velocity ÷ time taken
> $$a = \frac{v-u}{t}$$
> where $a$ is the acceleration of an object whose velocity changes from $u$ to $v$ in time $t$.
> Acceleration is measured in m/s².

The diagram, **Fig. 2.2**, shows a speed–time graph for part of a car journey.

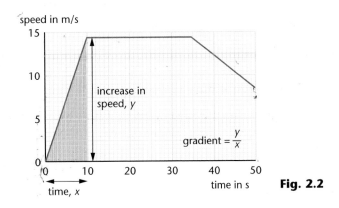

**Fig. 2.2**

# Force and motion

In this graph:

- the speed can only have positive values
- if the direction of motion does not change the acceleration can be calculated as *acceleration = change in speed ÷ time taken*; this is equivalent to calculating the gradient of the graph
- the steeper the gradient, the greater the acceleration that it represents
- a negative gradient shows a decrease in speed.

> **If the direction does not change, the change in speed is the same as the change in velocity.**

**KEY POINT** The gradient of a speed–time graph represents the acceleration in the direction of motion.

The graph also gives information about the distance that an object travels:

- distance travelled = average speed × time
- the distance travelled in the first 10 s is equal to $\frac{1}{2} \times 14.5 \times 10 = 72.5$ m; this is represented by the shaded area on the graph
- similarly, the distance travelled during the next 25 s is represented by the area of the rectangle between the graph line and the time axis.

> **The formula for distance travelled is derived from that for calculating speed.**

**KEY POINT** On a speed–time graph, the area between the graph line and the time axis represents the distance travelled.

A **velocity–time graph** differs from a speed–time graph because velocity can have both negative and positive values, showing motion in opposite directions.

On a velocity–time graph:

- if the gradient and velocity both have the same sign, the object is accelerating in the direction of motion
- if the gradient and velocity have opposite signs, the object is decelerating in the direction of motion
- the **total area** between the graph line and the time axis represents the distance travelled.

> **When working out the distance travelled, areas below the time axis count as positive; they are simply added on to areas above the axis.**

**PROGRESS CHECK**

1. Calculate the speed of the motion shown during the first 22 s in **Fig. 2.1**.
2. Calculate the acceleration of the motion shown in the first 10 s in **Fig. 2.2**.
3. Calculate the distance travelled during the 50 s of motion shown in **Fig. 2.2**.

1. 6.4 m/s;  2. 1.45 m/s²;  3. 607.5 m.

# 2.2 Movement and force

*After studying this section you should be able to:*

- describe how balanced and unbalanced forces affect the motion of an object
- recall and use the relationship between force, mass and acceleration
- explain how a falling object reaches a terminal velocity.

## Starting and stopping

Our everyday experience of motion tells us that things do not keep moving without a force. Remove the force and the motion eventually stops. Prior to the work of Galileo and Newton it was thought that there is only one force involved in motion – the **driving force**. Newton realised that there are "unseen" forces such as **friction** and **air resistance**. If you push a book across a table and let go it stops moving because the friction between the book and the table is an **unbalanced force**.

The size of the friction force depends on the roughness of the surfaces: the rougher the surfaces, the greater the friction force.

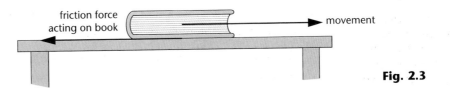

friction force acting on book → movement

**Fig. 2.3**

Forces due to friction:

- oppose slipping and sliding
- always act in the opposite direction of any motion
- cause **heating** and **wearing** of surfaces that rub together.

Friction is essential for walking, as well as starting and stopping the motion of a bicycle, bus or train.

Friction between a shoe and the ground prevents it from slipping. The force that propels a person who is walking is the forwards push of the ground on the shoe.

When you set off on a bike:

- the **wheel** pushes **backwards** on the road
- forces always act in pairs, so if object A pushes or pulls object B, then B pushes or pulls A with an equal-sized force in the opposite direction
- the **road** pushes **forwards** on the wheel, causing the bike to move.

**Fig. 2.4**

The forces between the wheel and the road are friction forces; without friction the wheel would just spin round.

Friction is also needed to stop the bike when **braking**. When the brakes are applied, friction between the brakes and the wheel rims causes the wheels to push forwards on the road. The resulting backward push of the road on the wheel brings the bike to a halt. If there is insufficient friction, the bike skids as it slides along the road surface.

## A question of balance

As you cycle along at a steady speed, resistive forces act. The main resistive force is **air resistance**. To maintain a constant speed in the same direction, the resistive forces need to be **balanced** by the driving force, so that equal-sized forces act both forwards and backwards.

Changing **speed** or **direction** requires an **unbalanced force** in the direction of the change:

- when speeding up the driving force is bigger than the resistive force
- when slowing down the resistive force is bigger than the driving force
- when turning a corner friction between the road and the wheels causes a sideways-force.

The diagram, **Fig. 2.5**, shows the **balance** of forces acting on a cyclist when speeding up, travelling at constant speed and braking.

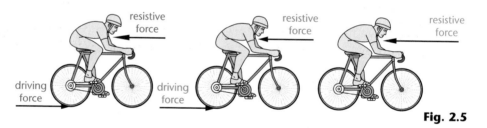

Fig. 2.5

## Force and acceleration

When the forces acting on an object are balanced, there is no change in its motion. It either remains stationary or moves at a constant velocity, ie there is no change in speed or direction. An **unbalanced force** causes a **change in velocity**, the object accelerates.

The acceleration caused by an unbalanced force:

- acts in the direction of the unbalanced force
- is proportional to the size of the unbalanced force
- is inversely proportional to the mass of the object

**Two quantities are proportional if doubling the size of one causes the other to double. They are inversely proportional if doubling the size of one causes the other to halve.**

> **KEY POINT**
>
> The relationship between the size of an unbalanced force, $F$, and the acceleration, $a$, it causes when acting on a mass, $m$, is
>
> force = mass × acceleration
>
> $$F = m \times a$$
>
> This relationship is used to fix the size of the unit of force, the newton (N), as the force needed to accelerate a mass of 1 kg at a rate of 1 m/s².

It is important to remember that this relationship applies to the size of the **unbalanced** force acting, rather than any single force. The size of the unbalanced force due to two forces acting along the same line is their sum if they are in the same direction and their difference if they act in opposite directions. In the diagram, **Fig. 2.6**, the size of the unbalanced force is 1000 N in the **forwards** direction.

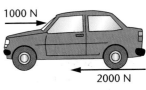

1000 N

2000 N

**Fig. 2.6**

## Stopping distance

The distance that a vehicle travels between the driver noticing a hazard and the vehicle stopping is known as the **stopping distance**:

● stopping distance consists of **thinking distance** and **braking distance**

● thinking distance is the distance travelled during the driver's reaction time, the time between noticing the hazard and applying the brakes

● braking distance is the distance travelled while the vehicle is braking.

The diagram, **Fig. 2.7**, shows how the stopping distance depends on vehicle speed.

**This diagram shows that thinking distance is proportional to vehicle speed and braking distance is proportional to the square of the vehicle speed.**

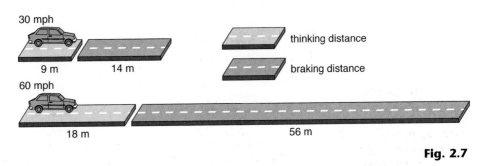

30 mph

9 m    14 m    thinking distance

braking distance

60 mph

18 m    56 m

**Fig. 2.7**

**Some medicines contain drugs that can make the user drowsy. They carry a warning that a person should not drive after taking the medicine.**

These distances could be greater:

● if the driver is tired or affected by any drugs or alcohol; the reaction time and the thinking distance are increased

● if the road is wet or icy or the tyres or brakes are in poor condition; the braking distance is increased

● if the vehicle is fully loaded; the extra mass reduces the deceleration during braking, so the braking distance is increased.

## Stopping quickly

In a head-on collision, vehicles can be brought to a halt in a very short time. The driver and passengers in a vehicle would **carry on moving** until they hit something that could cause sufficient force to decelerate them. This would probably be the car windscreen or the seat in front, and in either case the force on impact could cause severe injury.

**Seat belts** reduce the injury to drivers and passengers:

- seat belts prevent the occupants of a vehicle from hitting solid objects in front

- they stretch slightly so that the wearer is brought to rest over a **longer time** than by hitting the object in front

- this reduces the **deceleration** of the wearer, and the force required to stop him.

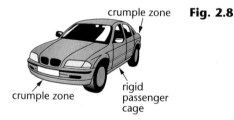

crumple zone **Fig. 2.8**

crumple zone

rigid passenger cage

Modern cars and trains are also designed with **crumple zones** at the front and rear. The vehicle takes longer to stop in a collision, reducing the deceleration and the force required to stop the passengers.

## Moving vertically

The vertical motion of an object is affected by the Earth's **gravitational field** which is responsible for the downward pull that is called weight. The size of this pull depends on the **gravitational field strength**, *g*. Close to the surface of the Earth this has a constant value of 10 N/kg, so that each kg of mass experiences a force of 10 N.

Gravitational field strength is equivalent to free-fall acceleration, the acceleration of an object falling in the absence of resistive forces. Their units, N/kg and m/s$^2$, are also equivalent.

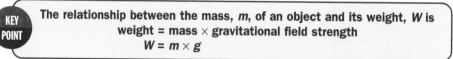

**KEY POINT**

The relationship between the mass, *m*, of an object and its weight, *W* is

weight = mass × gravitational field strength

$$W = m \times g$$

For a streamlined object moving vertically at low speeds the effect of air resistance is small, so it can be considered to have a constant acceleration equal to *g* and acting vertically downwards.

At greater speeds, and when the object is not streamlined, the effects of both its weight and air resistance have to be taken into account to explain its motion.

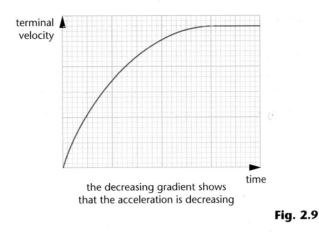

the decreasing gradient shows
that the acceleration is decreasing

**Fig. 2.9**

The graph, **Fig. 2.9**, shows how the vertical speed of a skydiver changes after she leaves an aircraft and before she opens her parachute:

- initially her acceleration is equal to *g* as there are no resistive forces when she is not moving

> **The unbalanced force is the difference between the gravitational force and the air resistance.**

- as her speed increases, so does the size of the air resistance

- this causes a decrease in the size of the unbalanced force pulling her down, and a consequent decrease in her acceleration

- when the size of the air resistance and her weight are equal she no longer accelerates as there is no unbalanced force

- the constant speed when the forces are balanced is called the terminal velocity.

---

**PROGRESS CHECK**

1. What is the relationship between the driving force and the resistive force when a cyclist is accelerating forwards?
2. A cyclist and his bike have a mass of 85 kg. What unbalanced force is needed to cause an acceleration of 1.2 m/s²?
3. Explain why the balance of forces on a skydiver changes as her speed increases.

1. The driving force is greater than the resistive force;  2. 102 N;  3. The gravitational force does not change but the air resistance increases as she speeds up.

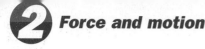 

# 2.3 The effects of forces

**LEARNING SUMMARY**

After studying this section you should be able to:

- describe the behaviour of a spring and a rubber band when subjected to an increasing stretching force
- apply the principle of moments to calculate the forces acting on a stable structure
- explain how the pressure exerted by a gas depends on its volume.

## Forces and materials

Edexcel A    Edexcel B
NICCEA
WJEC

Pulling or pushing on a material causes it to change its **shape**. Sometimes the change in shape is apparent, as in the stretching of a **rubber band**. On other occasions, such as when walking across a concrete floor, the change in shape is so minute as to be unnoticeable. Objects such as bridges are subjected to forces that **compress** some parts and **stretch** others. It is important that they are designed to withstand these changes in shape and not to undergo permanent deformation.

> When you walk across a floor or sit on a chair it compresses. This results in an upward force on you that balances your weight.

The diagrams, **Fig. 2.10**, show how the extension of a spring and a rubber band depends on the size of the stretching force.

In the case of the spring:

> The uniform pattern of extension of a spring is used in a spring balance or forcemeter, where equal increases in force result in equal increases in extension.

- the **extension** is **proportional** to the **force** up to the end of the straight line part of the graph, called the **limit of proportionality** (sometimes called the **elastic limit**)

- beyond this point the spring becomes harder to stretch, greater increases in force are required to cause the same increase in extension

- the spring may not return to its **original size** if stretched beyond the limit of proportionality.

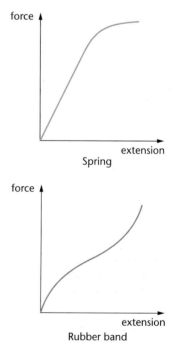

For the rubber band:

**Fig. 2.10**

> This does not happen if the force is large enough to break the rubber band.

- the extension is **not proportional** to the **stretching force**; the band becomes easier, then more difficult to stretch as the force is increased

- the band always returns to its **original size** and shape when the stretching force is removed.

## Turning forces

The **turning effect** of a force is used when opening a door, riding a bike and even when flushing the toilet. All these examples involve objects that are free to **rotate** around a fixed point or **pivot**. The effect that a force has in causing rotation depends on:

If the line of action of the force passes through the pivot, there is no turning effect.

- the size of the force
- the distance that it is applied from the pivot
- the angle at which the force is applied.

 **KEY POINT**
The moment, or turning effect of a force, is calculated using the relationship:
moment = force × perpendicular distance to pivot
The moment of a force is measured in N m.

When several turning forces act on an object, whether it turns round depends on the balance of the moments acting in an **anticlockwise** direction compared to those acting in a **clockwise** direction. If the sum of the moments in each direction is the same, then the object is balanced.

The principle of moments applies to the moments calculated about any single pivot.

**KEY POINT**
The principle of moments states that, when a system is balanced:
sum of clockwise moments = sum of anticlockwise moments

## Forces acting on a beam

When a heavy lorry travels across a beam such as a bridge across a road, the forces acting at the bridge supports change with the position of the lorry. The two rules used to calculate the size of these forces are:

- the total force acting upwards must equal the total force acting downwards
- the clockwise moment about each support must equal the anticlockwise moment.

The forces acting are shown in the diagram, **Fig. 2.11**. It is assumed that the weight of the bridge itself is small and can be ignored.

As the bridge travels from right to left, the force $R_1$ increases and $R_2$ decreases.

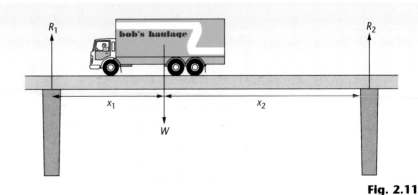

**Fig. 2.11**

Application of the above rules to this situation gives:

- $R_1 + R_2 = W$, since the sum of the upward forces equals the downward force
- $W \times x_1 = R_2 \times (x_1 + x_2)$, taking moments about the left hand support
- $W \times x_2 = R_1 \times (x_1 + x_2)$, taking moments about the right hand support.

## Gas pressure

Edexcel A   Edexcel B
NICCEA
WJEC

The **particles** of a gas are in constant motion. They are continually changing speed and direction, which is why their motion is often described as being "**random**". **Gas pressure** results from the **collisions** between the particles and their surroundings, including the walls of their container. The size of this pressure depends on:

- the **frequency** of the collisions
- the **mean speed** of the particles
- the **mass** of the particles.

> A common misconception at GCSE is that gas pressure is due to collisions between particles and other particles. This is not the case.

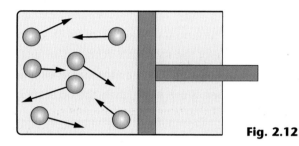

**Fig. 2.12**

Reducing the volume of a gas causes an increase in the frequency of the collisions between the gas particles and the container walls, resulting in an increased pressure.

> **KEY POINT**
> The relationship between the pressure and volume of a fixed mass of gas at constant temperature is:
> $$p \times V = \text{constant}$$
> $$\text{or } p_1 \times V_1 = p_2 \times V_2$$
> Where the subscripts 1 and 2 refer to the state of the gas before and after the change in pressure and volume.

> Pressure is measured in pascals (Pa), where 1 Pa = 1 N/m².

This relationship applies to a gas at a constant temperature. In practice, squashing a gas causes heating and a further increase in pressure due to the increased frequency of collisions and the increase in the mean speed of the particles. The reverse is true when a gas expands.

> **PROGRESS CHECK**
>
> 1. Why does a rubber band not have a "limit of proportionality"?
> 2. Explain why door handles are mounted as far from the hinge as possible.
> 3. The pressure of a gas is $4.0 \times 10^5$ Pa when its volume is $2.5 \times 10^{-3}$ m³. Calculate the new pressure when the volume is reduced to $1.2 \times 10^{-3}$ m³, assuming that there is no change in the temperature.
>
> 1. The extension is not proportional to the stretching force;   2. To maximise the moment and minimise the force needed to open the door;   3. $8.3 \times 10^5$ Pa.

# *Sample GCSE question*

**1.** A train accelerates from rest (0 m/s) to its maximum speed of 60 m/s in a time of 120 s.

**(a)** Calculate the acceleration of the train. **[3]**

> $Acceleration = change\ in\ velocity \div time\ taken$ ✓
> $= 60\ m/s \div 120\ s$ ✓
> $= 0.5\ m/s^2$ ✓

*Take care with the units of acceleration. A common error at GCSE is to give the unit as m/s instead of m/s².*

**(b)** The mass of the train is 350 000 kg.

**(i)** Calculate the size of the force needed to accelerate the train. **[3]**

> $force = mass \times acceleration$ ✓
> $= 350\ 000\ kg \times 0.5\ m/s^2$ ✓
> $= 175\ 000\ N$ ✓

*There is one mark here for recall of the formula; the second mark is for identifying the quantities correctly and the final mark is for obtaining the correct answer and giving the correct unit.*

**(ii)** Explain why the acceleration of the train decreases as its speed increases. **[3]**

> *Resistive forces act on the train* ✓.
> *These forces increase as the train's speed increases* ✓.
> *Reducing the size of the unbalanced force acting on the train* ✓.

*The size of the unbalanced force is the difference between that of the driving force and that of the resistive forces, since these act in opposite directions.*

**(c)** A jet aircraft of the same mass as the train accelerates from rest to a speed of 60 m/s before it takes off from the ground.

**(i)** Explain why the aircraft needs a much greater force than the train to reach the same speed. **[3]**

> *The aircraft needs to reach take-off speed in a shorter distance* ✓.
> *So it needs a greater acceleration* ✓.
> *A bigger force is needed to cause the greater acceleration* ✓.

*You are not expected to know anything about aircraft to answer this question. You are being asked to use your understanding of $f = m \times a$ to evaluate the information that you have been given.*

**(ii)** Suggest why it is important that aircraft do not take off with far more fuel than is needed for the journey. **[1]**

> *Excessive mass increases the distance required for take-off OR increases the fuel consumption on the journey* ✓.

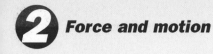 

# Exam practice questions

**1. (a)** The diagram shows the forces between the wheel of a bicycle and the road when the bicycle is accelerating forwards.

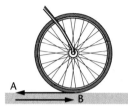

(i) Force A is the backwards push of the wheel on the road.

Write a similar description of force B. [2]

(ii) Which option describes the relative sizes of these forces correctly?

A Force A is greater than force B.

B The forces are equal in size.

C Force B is greater than force A.

Write down the letter of your choice. [1]

(iii) Explain why wet leaves between the wheel and the ground may cause the wheel to spin round. [2]

**(b)** The mass of the cycle and cyclist is 90 kg.

The forwards force has a value of 120 N.

The resistive force has a value of 60 N.

(i) What is the main resistive force that acts on the cyclist? [1]

(ii) Calculate the size of the unbalanced force on the cyclist and state its direction. [2]

(iii) Calculate the acceleration of the cyclist. [3]

(iv) Explain how the acceleration of the cyclist changes as the speed of the cycle increases. [3]

**2.** The graphs show how the thinking distance and braking distance of a car are related to its speed.

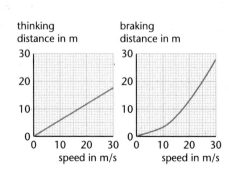

# Exam practice questions

**(a)** Describe the relationship between:

    **(i)** thinking distance and speed        **[1]**

    **(ii)** braking distance and speed.        **[1]**

**(b)** Use the graphs to give the values of the:

    **(i)** thinking distance at a speed of 26 m/s.

    **(ii)** braking distance at a speed of 26 m/s.

    **(iii)** stopping distance at a speed of 26 m/s.     **[3]**

**(c)** Write down TWO factors that could affect:

    **(i)** the thinking distance

    **(ii)** the braking distance of a vehicle.     **[2]**

**3.** The graph shows how the velocity of a ball changes after it is thrown vertically upwards.

**(a)** **(i)** Calculate the acceleration of the ball.     **[3]**

    **(ii)** What is the direction of the acceleration?

    Explain how you can tell.     **[2]**

    **(iii)** The mass of the ball is 0.020 kg.

    Calculate the size of the force required to cause this acceleration.     **[3]**

**(b)** **(i)** After what time shown on the graph did the ball change direction?

    Explain how you can tell.     **[2]**

    **(ii)** Use the graph to work out the height that the ball reached.     **[3]**

**4. (a)** Explain how gases exert pressure.     **[2]**

**(b)** Explain how the pressure exerted by a gas changes when its volume is decreased.     **[2]**

**(c)** A carbon dioxide cylinder contains gas at a pressure of $4.5 \times 10^5$ Pa.

The volume of the cylinder is 0.015 m³.

Calculate the volume occupied by the gas at atmospheric pressure, $1.0 \times 10^5$ Pa.

# Waves

**The following topics are covered in this section:**

- **Wave properties and sound**
- **Light and the electromagnetic spectrum**
- **The restless Earth**

# 3.1 Wave properties and sound

**LEARNING SUMMARY**

*After studying this section you should be able to:*

- *recall and use the wave equation*
- *explain how echoes are used to measure distance*
- *describe the effects of refraction and diffraction.*

## What is a wave?

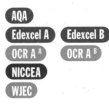

> If you write a letter, paper moves from you to the recipient, but this is not the case if you give the same information over the telephone.

> Waves that travel through the body of a liquid are longitudinal but surface water waves can be considered to be transverse.

A **wave** is a **vibration** or **oscillation**, a to-and-fro motion, which is transmitted through a material or through space. Waves can transfer **energy** and **information** from one place to another without the transfer of physical material.

There are some properties of waves that are common to different types of wave such as sound, light and radio waves. But there are also ways in which these waves behave differently.

Waves can be classified as either **longitudinal** or **transverse**, depending on the direction of the vibrations compared to the direction of wave travel:

- in a **longitudinal** wave, the vibrations are **parallel** to, or along, the direction of wave travel
- in a **transverse** wave, the vibrations are **perpendicular**, or at right angles, to the direction of wave travel
- sound waves are longitudinal; light and other electromagnetic waves (see section 3.2) are transverse.

The diagram, **Fig. 3.1**, shows the vibrations in a longitudinal and a transverse wave.

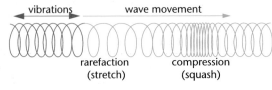

the vibrations in a logitudinal wave (above) and a transverse wave (below)

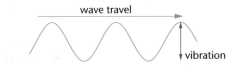

**Fig. 3.1**

# Wave properties

AQA
Edexcel A   Edexcel B
OCR A ᴬ   OCR A ᴮ
NICCEA
WJEC

## Wave measurements and the wave equation

These measurements apply to all waves:

- the **amplitude** (symbol $a$) of a wave motion is the greatest displacement (change in position) from the rest position

> A common error at GCSE is to describe the amplitude of a transverse wave as the distance from the top of a peak to the bottom of a trough; it is actually half that distance.

- **wavelength** (symbol $\lambda$) is the length of one complete cycle of a wave motion – a squash and a stretch for a longitudinal wave, a peak and a trough for a transverse wave

> The relationship between the frequency, *f*, and the time for one oscillation, *T*, is *f* = 1/*T*.

- the **frequency** of a wave (symbol $f$) is the number of vibrations each second; frequency is measured in hertz (Hz).

Increasing the **frequency** of a wave causes a decrease in the **wavelength**. High-frequency waves have short wavelengths, and low-frequency waves have long wavelengths. Frequency and wavelength are related to wave speed by the **wave equation**.

> **KEY POINT**
>
> For all waves, the relationship between wavelength and frequency is:
> **speed = frequency × wavelength**
> $$v = f \times \lambda$$

The **pitch** and **loudness** of a sound are determined by the frequency and amplitude of the wave:

> Because of the way in which our ears respond to sounds, changing the frequency of a wave can also change the perceived loudness, even though there is no change in amplitude.

- increasing the **amplitude** increases the **loudness** of a sound; a high-amplitude wave sounds louder than one with a lower amplitude

- increasing the **frequency** increases the **pitch** of a sound; a high-frequency sound wave has a higher pitch than a low-frequency wave

- humans can detect sound within the frequency range 20 Hz to 20 000 Hz, but the upper limit is reduced with increasing age

- **compression** waves above the maximum frequency that humans can detect are called **ultrasound**.

## Wave reflections and echoes

Echoes of sound and images in mirrors are due to the **reflection** of waves at a surface. **Fig. 3.2** shows the reflection of water waves in a ripple tank.

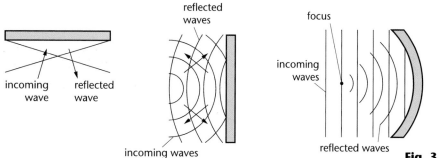

**Fig. 3.2**

These diagrams in **Fig. 3.2** show that:

- **plane** (straight) **waves** bounce off a barrier at the same angle as they hit it, or "the **angle of incidence** is equal to the **angle of reflection**"

- **circular** waves from a point source are reflected as if they came from a point behind the barrier; this point is the position of the **image**, the point where the reflected waves appear to have come from

- when plane waves are reflected at a concave barrier they are brought to a **focus**; this is what happens in a satellite dish.

> When you look at your own image in a mirror, it appears to be behind the mirror (see section 3.2).

Reflections of sound are called **echoes**. Echoes of sound and ultrasound are used for measuring distances and producing images of the inside of the body.

The diagram, **Fig. 3.3**, shows ultrasound being used to measure the depth of the sea bed.

When used to measure distance:

- a pulse of ultrasound is emitted from a vibrating crystal

- the same crystal then detects the reflected pulse

- since the pulse has travelled from the source, to the object and back again, the distance is calculated as $\frac{1}{2} \times$ speed $\times$ time.

> The pulse of ultrasound travels twice the distance between the source and the object being detected, since it goes there and back again.

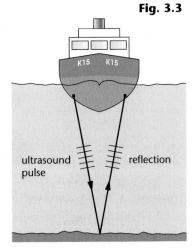

**Fig. 3.3**

ultrasound pulse    reflection

Ultrasound is also reflected at boundaries between layers of different materials and body tissue. **Ultrasound scans** are used to examine railway lines to detect cracks. They are used by medical staff to look at internal organs and fetuses without the need for any incision, so there are no scars to heal. When examining delicate organs and fetuses, an ultrasound scan is safer than an X-ray picture because ultrasound causes no damage to body cells or DNA.

## Refraction

Waves change speed when they travel from one material into another or when there is a change in density of the material that they are travelling in. This change of speed causes a change in direction. The change in direction is called **refraction**. The diagram, **Fig. 3.4**, shows the refraction of water waves in a ripple tank.

> In these diagrams, the waves slow down as they pass from deep water into shallow water.

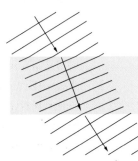

Using a ripple tank    Refraction causes waves to change direction  **Fig. 3.4**

When waves are **refracted**:

- the reduction in speed causes a corresponding reduction in wavelength

- the frequency of the waves does not change

- the direction of wave travel does not change if it is at right angles to the boundary between the surfaces

- if the waves meet the boundary at any other angle there is a change in direction.

> A virtual image is one that light does not pass through and cannot be projected onto a screen. Mirrors form virtual images.

A swimming pool always looks to be shallower than it really is. This is because the refraction of light as it crosses the water-air boundary causes a **virtual image** of the swimming pool floor to be formed. The diagram, **Fig. 3.5**, shows how the change in direction of light crossing this boundary deceives the eye about the position of the swimming pool floor.

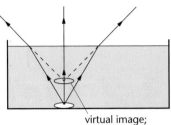

**Fig. 3.5**

virtual image; this is where the light appears to have come from

## Diffraction

> When answering questions about diffraction, it is important to emphasise that the amount of spreading of a wave depends on the size of the gap compared to the wavelength.

The spreading out of waves when they pass an obstacle or through a gap is called **diffraction**. Diffraction explains how **sound** can be heard around a corner and spreads along a corridor through an open doorway. Diffraction of **light** is less observable than that of sound because of the vast difference in their wavelengths. The amount of spreading of a wave when it is diffracted depends on the relative sizes of

**Fig. 3.6**

the gap that the wave passes through and the wavelength of the waves. This is shown in the diagram, **Fig. 3.6**.

These diagrams shows that:

- there is no detectable diffraction when the gap is many wavelengths wide, as in diagram **a**

- there is some spreading of a wave after it passes through a gap that is several wavelengths wide, as in diagram **b**

- after passing through a gap which is one wavelength wide, as in diagram **c**, the maximum spreading of the wave occurs, it appears to originate from the gap.

If you throw a tennis ball against a flat wall, it bounces off at the same angle as it hits. If you try to throw it through a gap, it either goes through or it doesn't, depending on whether the gap is larger or smaller than the ball (and your aim!). It does not spread out after passing through the gap. Both particles and waves show similar behaviour when they are reflected at a flat surface but particles cannot be diffracted. The fact that sound and light can both be diffracted shows that they have a wave-like behaviour.

**PROGRESS CHECK**

1. Sound travels in air at a speed of 330 m/s.
   Calculate the frequency of a sound that has a wavelength of 0.50 m.
2. SONAR uses sound to measure distances under the surface of the sea.
   A submarine sends out a sound pulse and receives the echo from a ship after 4.50 s.
   The speed of sound in sea water is 1500 m/s.
   How far away is the ship?
3. What is the condition for a wave to spread out in all directions after passing through a gap?

1. 660 Hz;   2. 3375 m;   3. The gap should be the same size as the wavelength of the waves.

# 3.2 Light and the electro-magnetic spectrum

**LEARNING SUMMARY**

*After studying this section you should be able to:*

- *explain how the reflection and refraction of light can lead to image formation*
- *describe how total internal reflection is used in reflecting prisms and optical fibres*
- *recall the main parts of the electromagnetic spectrum and their uses.*

## Images from light

AQA
Edexcel A   Edexcel B
OCR A $^A$   OCR A $^B$
NICCEA
WJEC

When light is **reflected** at a mirror, or partially reflected by a sheet of glass or a flat water surface, an image is formed. This image is:

> When light is reflected by a smooth, flat surface, it bounces off at the same angle as it hits.

- **virtual** – it does not exist but is seen where the brain "thinks" it is

- **directly behind the mirror**, the same distance behind the mirror as the object is in front

- **upright** (not inverted)

- the **same size** as the object.

An image is seen because the eye-brain system assumes that light travels in **straight lines**, and uses this principle to determine where the light has come from. This is shown in the diagram, **Fig. 3.7**.

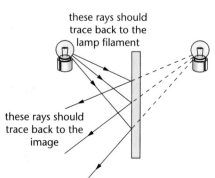

these rays should trace back to the lamp filament

these rays should trace back to the image

**Fig. 3.7**

Light also forms virtual images when it is **refracted**, changing speed and direction as it passes from one material into another. Light is slowed down when it passes into glass or perspex, and it speeds up again as it emerges. The changes in direction caused by the changes in speed fool the eye-brain system when it works out where the light has come from, so we "see" objects as being closer than they really are.

When light is refracted:

● there is always some reflection at the boundary between two materials

● any change in direction is **towards** a line drawn at right angles to the surface (this is called a **normal** line) when light slows down, and away from this line when light speeds up

● light emerges from a parallel-sided block **parallel** to, but displaced from, the direction in which it entered.

The change in direction of light passing through a glass block and the consequent image formation are shown in the diagrams, **Fig. 3.8**.

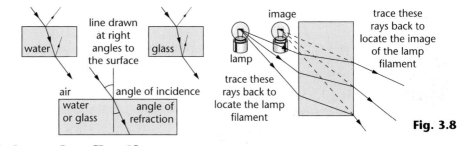

Fig. 3.8

## Internal reflection

Light does not always pass between two transparent materials when it meets the boundary between them. In cases where light would speed up as it crosses a boundary it is possible for all the light to be reflected. The proportions of the light that are **reflected** and **refracted** depends on the **angle of incidence**, the angle between the light that meets the boundary and the **normal** line. The behaviour of light meeting a glass-air boundary at different angles of incidence is shown in the diagram, **Fig. 3.9**.

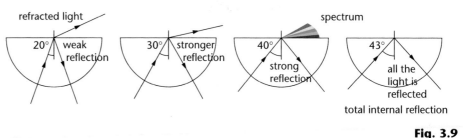

Fig. 3.9

When light meets a glass-air boundary:

● at a small angle of incidence, some light is reflected and some is refracted

● as the angle of incidence is increased, a greater proportion of the light is reflected

● at the **critical angle**, about 42°, the light that leaves the glass is parallel to the surface

● at angles of incidence greater than the critical angle, **total internal reflection** takes place; no light leaves the glass, it is all reflected internally.

Total internal reflection is used in reflecting **prisms** and to transmit data along **optical fibres**.

**Reflecting prisms are preferred to mirrors in periscopes because there is no silvered surface that can deteriorate and they absorb less of the light.**

Reflecting prisms are used in periscopes, binoculars, cycle reflectors and some cameras. The reflecting prisms in periscopes turn the light round a 90° corner; in binoculars and cycle reflectors the light undergoes two reflections, resulting in a 180° change in direction. This is shown in the diagram, **Fig. 3.10**. In each case the internal angles of the prism are 45°, 45° and 90° so that whenever light meets a glass-air surface the angle of incidence is 45°, which is **greater** than the **critical angle**.

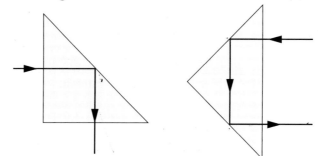

**Fig. 3.10**

**One advantage of optical fibres is that light can travel along them with very little energy loss.**

Light can travel "round the bend" in an **optical fibre** by repeated reflection at the glass-air boundary. Provided that when light meets this boundary the angle of incidence is greater than the critical angle, all of the light is **reflected internally** and none passes out of the fibre. This is shown in the diagram, **Fig. 3.11**.

An **endoscope** uses two bundles of fibres to see inside the body of a patient. One bundle carries light into the body to illuminate the area being examined. A second bundle carries the reflected light to a television camera, which produces an image on a television screen.

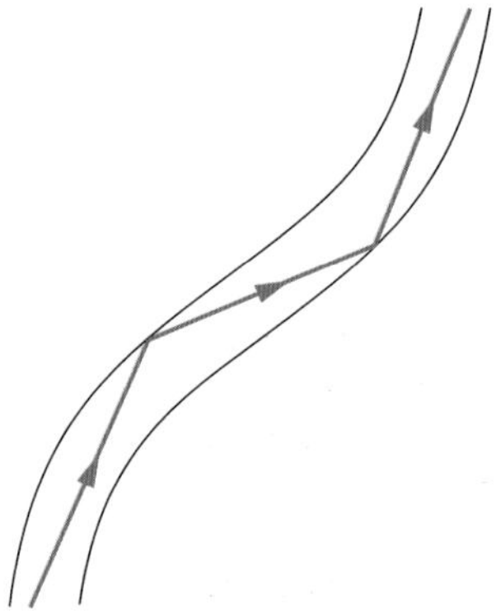

**Fig. 3.11**

## A spectrum of waves

AQA
Edexcel A Edexcel B
OCR A[A] OCR A[B]
NICCEA
WJEC

**The range of wavelengths of light in air is approximately $4 \times 10^{-7}$ m (blue) to $6.5 \times 10^{-7}$ m (red).**

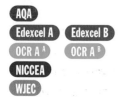

White light consists of waves with a range of **wavelengths** and **frequencies**. It is split up into the colours of the rainbow when it passes through water droplets. A glass or perspex **prism** can have a similar effect:

- when light passes into the prism, light at the blue (short wavelength) end of the spectrum undergoes a greater change in speed than light at the red (long wavelength) end of the spectrum
- the change in direction depends on the change in speed

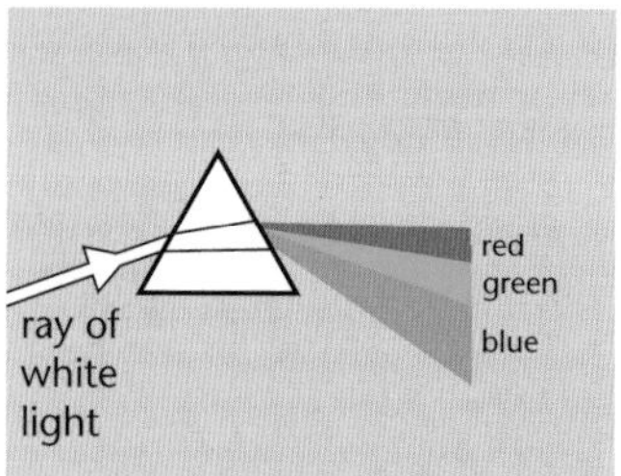

red
green
blue
ray of white light


**Fig. 3.12**

- a similar effect occurs when light leaves the prism
- this splitting up of white light into colours is called **dispersion**.

The dispersion of white light by a prism is shown in the diagram, **Fig. 3.12**.

Visible light is only a small part of a whole family of waves with similar properties. All these waves are called **electromagnetic**; this describes the type of oscillation that makes the wave motion. The whole family is collectively referred to as the **electromagnetic spectrum**. It ranges from the shortest electromagnetic waves, **X-rays** and **gamma rays**, to the longest, **radio waves**. The diagram, **Fig. 3.13**, shows the range of wavelengths and frequencies of the different waves that make up the spectrum.

**Different types of electromagnetic waves are produced in different ways and have different effects.**

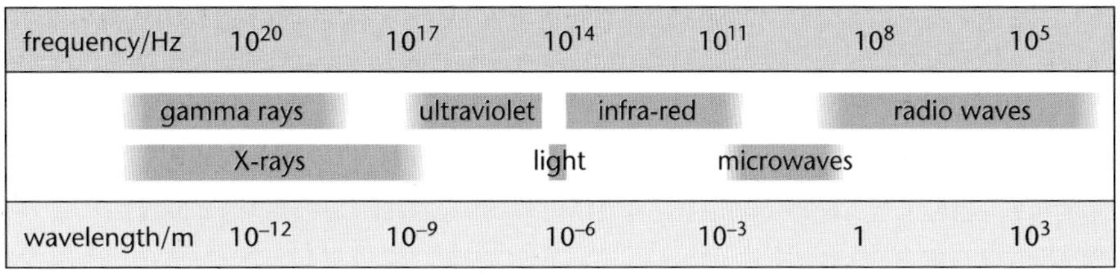

| frequency/Hz | $10^{20}$ | | $10^{17}$ | $10^{14}$ | $10^{11}$ | $10^{8}$ | $10^{5}$ |
|---|---|---|---|---|---|---|---|
| | | gamma rays | | ultraviolet | infra-red | | radio waves |
| | | X-rays | | light | | microwaves | |
| wavelength/m | $10^{-12}$ | | $10^{-9}$ | $10^{-6}$ | $10^{-3}$ | 1 | $10^{3}$ |

**Fig. 3.13**

All electromagnetic waves:

- travel at the same speed in a vacuum, $3.0 \times 10^8$ m/s
- transfer energy and cause heating when they are absorbed.

## The shortest waves

**X-rays** and **gamma rays** have the shortest wavelengths and carry the most energy. They are also the most **penetrative** of the electromagnetic waves, a property which is useful in medical imaging. When an X-ray photograph is taken:

**X-rays and gamma rays are produced in different ways. X-rays come from X-ray tubes and gamma rays are emitted by unstable nuclei. There is no difference in the waves.**

- X-rays are passed through the body and detected by a **photographic plate**
- the X-rays pass through the flesh and are absorbed by the bone
- bone shows up as white on the photograph and flesh appears dark
- a bone fracture is seen as a dark line on the white bone in an X-ray photograph.

A similar technique is used to examine the turbine blades of jet engines, to check for cracks.

**Gamma rays** are useful for checking individual organs when used as a **tracer** (see section 6.2). A radioactive isotope that emits gamma rays is injected into the body and when it has circulated it can be detected by a camera to give either a still or a moving picture. The radioactive isotopes can be made so that they concentrate in particular areas of the body.

Although they are only weakly absorbed by body tissue, X-rays and gamma rays are both **ionising** radiations. They can destroy cells and cause **mutations** in the DNA. Because of this:

- both are used in the treatment of cancer to destroy abnormal cells
- gamma rays are used to kill bacteria in food and to sterilise medical instruments
- people who come into contact with X-rays and gamma rays need to be protected from damaging over-exposure.

# Either side of light

**Ultraviolet** radiation is **higher energy** and shorter wavelength than light. Much of the ultraviolet radiation from the Sun is absorbed by the atmosphere, but in the summer months less is absorbed than in winter. Ultraviolet radiation is also produced when an electric current passes in a tube containing mercury vapour. These tubes are used in sunbeds and in fluorescent lights, where a coating on the inside of the tube absorbs the ultraviolet radiation and re-emits it as light. Security pens use a fluorescent paint which is hardly visible in normal lighting but glows brightly when illuminated with an ultraviolet lamp.

> **Because of the Earth's tilt, radiation from the Sun passes through more of the Earth's atmosphere in winter than in summer.**

In addition to these uses, ultraviolet radiation can be harmful to humans:

- absorption by the **skin** can cause **cancer**

- light skins are more prone to cancer than dark skins, as they allow the radiation to penetrate further into the body

- absorption by the **retina** can cause **blindness**.

> **It is important to protect the skin and eyes from ultraviolet radiation when outside in the summer months.**

**Infra-red** radiation has a longer wavelength and **lower energy** than light, so it is less harmful than ultraviolet radiation. All objects **emit** infra-red radiation and **absorb** it from their surroundings. The **hotter** the object, the **greater** the rate of emission. Infra-red radiation is used:

- in cooking; toasters and grills transfer energy to food by infra-red radiation

- in night-time photography; people and animals can be distinguished from their surroundings because they emit more infra-red radiation

- to find and rescue people trapped in the rubble of buildings after an earthquake

- in remote controls for devices such as televisions and hi-fi.

Over-exposure to infra-red radiation from the Sun causes **sunburn**.

The **microwaves** used in cooking have a wavelength around 12 cm, so they fit into the **radio wave** part of the electromagnetic spectrum. Radio waves of this wavelength can penetrate a few centimetres into food, and they have the right frequency to be **absorbed** by **water molecules** as they pass through. Food becomes cooked by the following process:

> **You cannot tune in a radio to a microwave oven as the waves do not carry a signal.**

- water molecules absorb the energy of the microwaves, increasing their energy of **vibration**

- this energy is transferred to other molecules in the food by **conduction**

- food is cooked uniformly as this process takes place throughout the body of the food.

# Communicating with electromagnetic waves

National television and radio programmes are broadcast using **radio waves**. Television broadcasts from your local transmitter to your aerial use wavelengths around 0.6 m; the wavelengths used for radio range from around 3 m for VHF to hundreds of metres for medium and long-wave broadcasts. Long wavelength radio waves:

- are **diffracted** more around hills and buildings, so there is less chance of a "shadow" causing poor reception

- have a **low frequency** which limits the amount of information that can be transmitted.

The signal is carried by changes in either the amplitude or frequency of the radio wave. This is known as amplitude modulation (AM) or frequency modulation (FM).

Much shorter wavelength waves, **microwaves**, are used to send the information from London to regional transmitters. These radio waves travel as a narrow beam and so short-wavelength radio waves are used to minimise the effects of **diffraction**. The wavelengths used are typically a few centimetres.

Diffraction effects are even more important when communicating with satellites. The waves used here have wavelengths of a few millimetres. An ordinary television set cannot detect these wavelengths, so to receive satellite television transmissions you also need a set-top box which takes the information from the microwaves and transfers it to longer wavelength waves that can be interpreted by the television set.

The signals received from satellites are very weak, so a dish aerial is used to gather the energy reaching a wide area.

Television, radio and telephone conversations can be sent as either **analogue** or **digital** signals. The amplitude or frequency of an analogue signal varies continuously with time but a digital signal can only be in one of two states – either "on" or "off". The diagram, **Fig. 3.14**, shows a digital and an analogue signal.

Both analogue and digital signals become **distorted** as they travel in any medium. Digital signals allow higher quality transmission because:

Digital signals are increasingly being used for all types of communications.

- a digital signal can be returned to its original state in a process known as **regeneration**

- this is possible because any noise or distortion can be detected and removed

- in an analogue signal there is no way of distinguishing between noise and the original signal, and when the signal is amplified so is the noise.

an analogue signal

a digital signal

**Fig. 3.14**

Digital signals are also used to transmit data along **optical fibres**. Light has a higher frequency than radio waves, allowing more information to be carried. The diagram, **Fig. 3.15**, shows how optical fibres are used in telephone links.

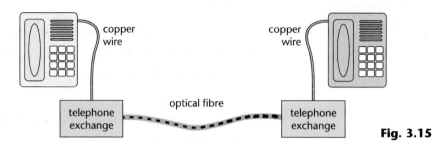

Fig. 3.15

**PROGRESS CHECK**

1. What causes the change in direction that can occur when light passes from one material into another?
2. Which three types of wave in the electromagnetic spectrum have wavelengths shorter than light?
3. Why are microwaves used for satellite communication?

1. The change in speed;   2. Ultraviolet radiation, X-rays and gamma rays;   3. To minimise the effects of diffraction.

# 3.3 The restless Earth

**LEARNING SUMMARY**

*After studying this section you should be able to:*

- *explain how evidence for the structure of the Earth comes from the waves detected after an earthquake*
- *describe the structure of the Earth*
- *explain how movement of plates in the lithosphere results in the recycling of rock.*

## Evidence for the Earth's structure

**Earthquakes** produce three types of wave that travel through the Earth and are detected by instruments called **seismometers**. The diagram, **Fig. 3.16**, shows a recording of the waves detected by a seismometer.

A seismometer recording is called a seismographic record.

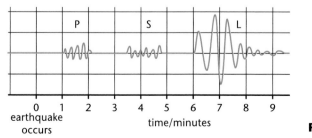

Fig. 3.16

The relative speeds of the waves can be deduced from the seismometer recording shown in Fig. 3.16.

The waves detected are:

- **L waves**; these are **long-wavelength** waves that travel around the Earth's crust. They are responsible for damage to buildings caused by movements in the ground.

- **P waves** or primary waves; these are **longitudinal** waves that can travel through both solids and liquids. They have the greatest speed of the waves caused by an earthquake and are the first to be detected.

- **S waves** or secondary waves; these **transverse** waves can only travel through solid materials. They are detected after the primary waves because of their lower speed.

The diagram, **Fig. 3.17**, shows where P and S waves are detected following an earthquake.

This diagram shows that:

- both P and S-waves follow curved paths due to **refraction** as the speed of the waves increases due to increasing density of the material that they are travelling through

- S waves are not detected in the **shadow region** on the opposite side of the Earth to the centre of the earthquake, called the **epicentre**

- there is a change in direction when the P waves cross a boundary, caused by the change of speed due to a change in density.

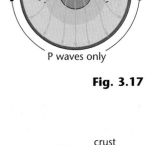

**Fig. 3.17**

A transverse mechanical wave can be transmitted along the surface of a liquid, but not through its bulk.

> **KEY POINT**
> The fact that S waves form a shadow region on the opposite side of the Earth shows that part of the core is in a liquid state.

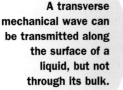

**Fig. 3.18**

This provides evidence for the structure of the Earth shown in **Fig. 3.18**:

- a thin **outer crust** (**lithosphere**) whose thickness varies between 10 km (under oceans) and 65 km (under mountains)

- a **mantle** that behaves like a solid but allows very slow convection currents to transfer energy from the centre to the surface

- a metallic **core**, consisting mainly of nickel and iron

- the outer core is liquid but the inner core, although hotter, is solid due to the intense pressure on it.

## Movement in the Earth's crust

AQA
Edexcel A   Edexcel B
OCR A ᴬ   OCR A ᴮ

Energy is being released in the core of the Earth due to **radioactive decay**. This results in a net energy flow from the core to the surface. The energy is carried through the mantle by slow-moving **convection currents**. These are thought to be responsible for the movement in the Earth's crust that has caused the continents to form from a single land mass.

> The mantle is usually considered to be solid, but there is a slow movement of hot rock through it.

The diagram, **Fig. 3.19**, shows how the shapes of the continents fit together, giving evidence that they could once have been joined.

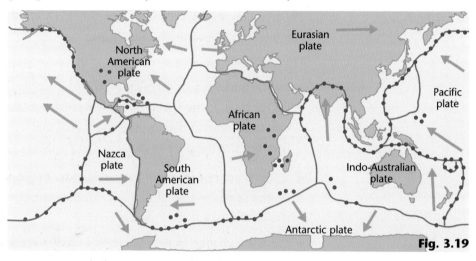

Fig. 3.19

The Earth's crust consists of a number of sections called **tectonic plates**. These are moving at the rate of a few centimetres each year. The plates can:

- slide past each other

- move towards each other

- move away from each other.

> The theory of plate tectonics was first proposed by Alfred Wegener in 1912. However, it was not universally accepted until the 1960s, when evidence of the spreading of the sea floor became available.

When plates slide past each other, it is not a smooth movement. **Resistive forces** between the plates prevent motion until the internal forces due to **compression** and **stretching** of the material is the greater force. This results in a sudden break in the ground as the plates move with a "jerk". This movement has caused the San Andreas Fault in California.

When plates move towards each other an ocean floor plate is forced under a continental plate in a process called **subduction**. When this occurs:

- the **ocean floor plate** partially melts due to the high temperatures in the **magma**

- **metamorphic rocks** form due to the **recrystallisation** of sedimentary rocks subjected to increased temperature and pressure

- mountains can be formed due to **folds** in the continental plate; this is how the Andes mountains in South America were formed

- these mountains can be **volcanic** due to **magma** rising through faults in the crust, resulting in the formation of fine-grained rocks due to rapid cooling of the magma.

Cracks under the sea floor are due to plates moving away from each other. This is a **constructive plate margin**, as it results in the formation of new rock:

● material from the **magma** flows out of the cracks and solidifies, producing new rock

● the new rocks formed are rich in iron and become **magnetised** by the Earth's magnetic field as they solidify

● the rocks in the ocean contain a record of the direction of the Earth's magnetic field; this record shows that it **reverses** once every few thousand years.

The diagram, **Fig. 3.20**, shows **subduction** at South America and the formation of a **magnetic record** at a mid-ocean ridge.

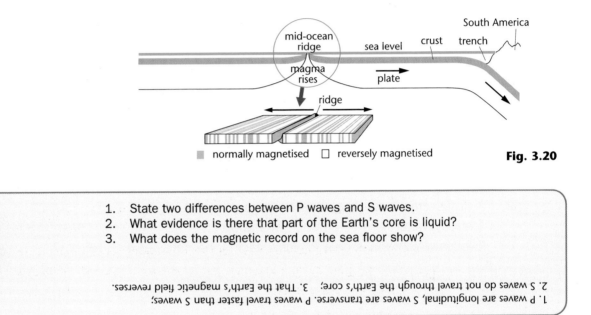

**Fig. 3.20**

**PROGRESS CHECK**

1. State two differences between P waves and S waves.
2. What evidence is there that part of the Earth's core is liquid?
3. What does the magnetic record on the sea floor show?

1. P waves are longitudinal, S waves are transverse. P waves travel faster than S waves; 2. S waves do not travel through the Earth's core; 3. That the Earth's magnetic field reverses.

# Sample GCSE question

**1.** Loudspeakers reproduce sounds which are transmitted through the air to the ears.

**(a)** Describe how sound travels through the air. **[2]**

> *It travels as a longitudinal wave ✓ in which*
> *the air particles vibrate parallel to the direction of wave*
> *travel ✓.*

*The answer here states how the sound travels (as a longitudinal wave) and then gives a description of the motion that forms the wave to gain 2 marks.*

**(b)** A high-pitched note has a frequency of 3500 Hz.
This travels through the air at a speed of 330 m/s.
Calculate the wavelength of the wave. **[3]**

> $\lambda = v/f$ *or wavelength = speed ÷ frequency* ✓
> $= 330 \text{ m/s} \div 3500 \text{ Hz}$ ✓
> $= 0.094 \text{ m}$ ✓

*The first mark is for recall and transposition of the wave equation.*
*Full marks are always awarded for a correct numerical answer with the correct unit. Only two marks would be awarded here if the unit was missing or wrong.*

**(c)** **(i)** Sound is diffracted as it leaves a loudspeaker.
Explain what this means. **[2]**

> *The sound spreads out ✓ as it passes through the*
> *opening at the front of the loudspeaker ✓.*

**(ii)** A hi-fi system has three separate loudspeakers.
Their diameters are 10 cm, 20 cm and 30 cm.
Explain which one is best for reproducing the note in **(b)**. **[4]**

> *The 10 cm loudspeaker is best ✓. The loudspeaker*
> *diameter is comparable to the wavelength ✓ so*
> *the sound will spread out as it leaves the loudspeaker ✓.*
> *The other loudspeakers have diameters greater*
> *than the wavelength, so the sound would not spread*
> *sufficiently to be heard anywhere in the room ✓.*

**(d)** The diagram shows how light passes through an opening of width 10 cm. Explain why light is not diffracted in the same way as sound at a similar size opening. **[3]**

*When answering questions about diffraction, it is important to stress that the amount of spreading depends on the size of the gap compared to the wavelength.*

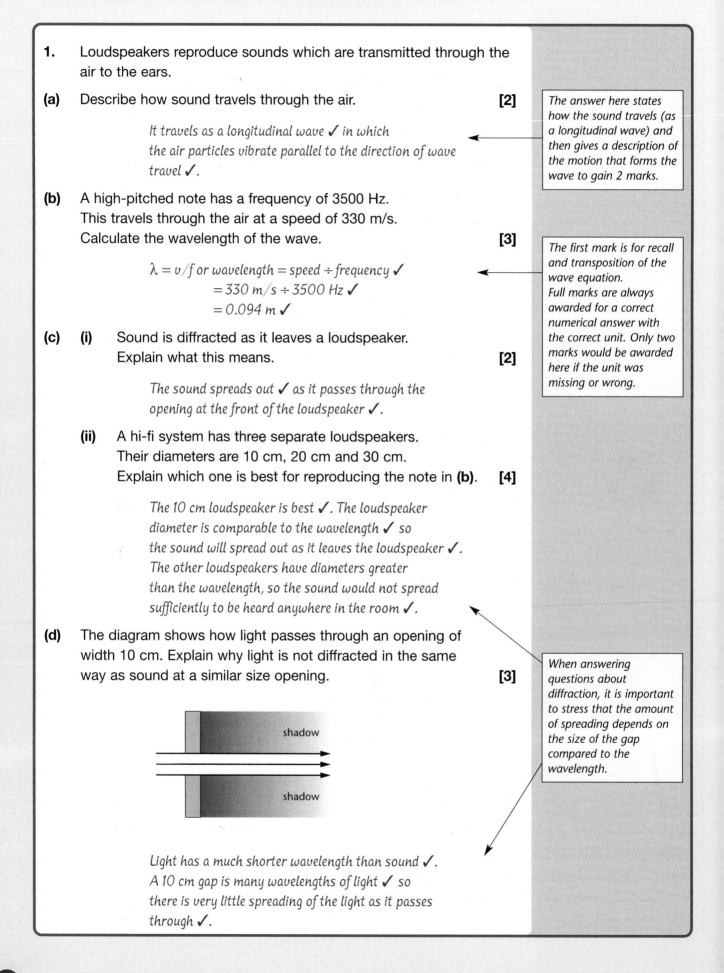

> *Light has a much shorter wavelength than sound ✓.*
> *A 10 cm gap is many wavelengths of light ✓ so*
> *there is very little spreading of the light as it passes*
> *through ✓.*

# Exam practice questions

**1.** Here is a list of some waves:

infra-red    sound    light    radio    ultraviolet    gamma    ultrasound

**(a)** Write down two waves from the list that are transverse. **[2]**

**(b)** Write down one wave from the list that can cause fluorescence. **[1]**

**(c)** Which of the waves in the list is used in an electric toaster to toast bread? **[1]**

**(d)** Write down one wave in the list that cannot be transmitted through a vacuum. **[1]**

**(e)** Pulses of sound or ultrasound can be used to measure distances.

The diagram shows how a house surveyor uses a sonic measuring device.

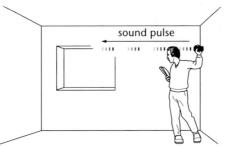

A pulse of sound is emitted and a short time later the width of the room is displayed on a screen.

**(i)** Suggest how the device uses the sound pulse to measure the width of the room. **[3]**

**(ii)** Explain how the device could give unreliable results if used in a room full of people or furniture. **[2]**

**2.** The diagram shows how water waves spread out after passing through a gap.

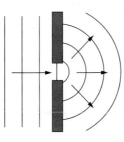

**(a)** Write down the name of this effect. **[1]**

**(b)** What two factors determine the amount by which the wave spreads out? **[2]**

**(c)** A typical sound wave has a wavelength of 1 m.
A typical light wave has a wavelength of $5.0 \times 10^{-7}$ m.
Explain why sound spreads out after passing through a doorway but light does not. **[2]**

# Exam practice questions

3.  When there is an earthquake, both longitudinal waves (P waves) and transverse waves (S waves), travel through the Earth.

    These waves are detected all over the Earth's surface by instruments called seismometers.

    The diagram shows the layered structure of the Earth.

    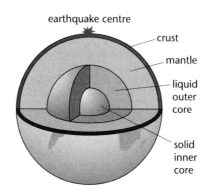

    (a)  The next diagram shows part of a seismometer record from a place close to the centre of the earthquake.

    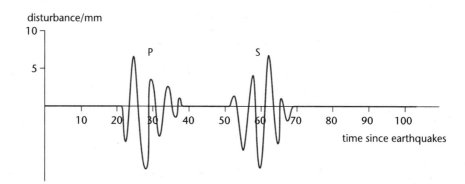

    (i)   Which type of wave travels faster, P waves or S waves?
          Explain how you can tell.                                                    [2]

    (ii)  Explain how the record from a seismometer placed on the Earth's surface directly opposite the centre of the earthquake would differ from that shown.                                                    [2]

    (b)  Use the top diagram to explain how seismometer readings give evidence that the outer core is liquid.                                                    [4]

## Satellites

**Artificial satellites** around the Earth are used for navigation, surveillance, communication, seeing into space and monitoring the weather. Many communications satellites, used for telephones and television transmission, occupy **geostationary orbits**. A satellite in a geostationary orbit:

● is directly above the equator

● has an orbit time of 24 hours

● stays above the same point on the Earth's surface.

Because all geostationary orbits have the same height, there is a limit to the number that can occupy this orbit at any one time.

Satellites such as weather satellites that are used to monitor the Earth are placed in **low polar orbits**. These have an orbit time of one and a half hours, so they orbit the Earth sixteen times each day, seeing a slightly different view of the Earth on each orbit, as the Earth spins on its axis.

## Is there life beyond the Solar System?

**In astronomy, the billion is used in the American sense of meaning one thousand million.**

Our Sun is one of the two hundred billion stars that make up the **Milky Way** galaxy, a collection of **stars** held together by gravitational forces. The Milky Way is a **spiral** galaxy, the stars forming the shape of a number of arms that spiral out from a central bulge. The galaxy is so vast that it takes light one hundred thousand years to pass between its extreme edges. It is rotating round at high speed, but the distances are so great that it takes two hundred million years for the Sun to complete one rotation.

There are more than a thousand billion known **galaxies** in the **Universe**. We know that life exists on one small star in our own galaxy. With so many galaxies and stars many people believe that the "chance" that created life on our planet must also have created life on other similar planets in the Universe. To try to detect such life:

**The search for extra-terrestial intelligence is known as SETI.**

● robots can be sent to nearby planets such as Mars to search for evidence of **microbes** or their fossilised remains

● the **atmospheres** of distant planets can be analysed, by detecting the different wavelengths of light that they transmit, to see if there is oxygen-enrichment due to plant life

● **radio telescopes** are used to search for radio signals from other advanced species of animal.

No evidence for any other life form existing has yet been found, but the immense size of the Solar System and the Universe means that direct exploration is very limited, and evidence in the form of radio waves could take many millions of years to reach us.

1. Between which two planets is the asteroid belt?
2. The Sun's gravitational pull on Jupiter, the fifth planet out from the Sun, is greater than that on the Earth, the third planet. Suggest why.
3. What is the advantage of a weather satellite occupying a low polar orbit rather than a geostationary orbit?

1. Mars and Jupiter;   2. Jupiter is very massive;   3. In a low polar orbit, the satellite can monitor the whole of the Earth in one day.

# 4.2 Evolution

**LEARNING SUMMARY**

*After studying this section you should be able to:*

● *describe the life cycles of stars*
● *explain how movement of the galaxies supports the "Big Bang" theory*
● *explain how the future of the Universe depends on the amount of mass contained within it.*

## The life of a star

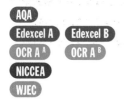

AQA
Edexcel A    Edexcel B
OCR A ^A    OCR A ^B
NICCEA
WJEC

**Nuclear fusion is the process in which nuclei join together. Very high speeds are needed for this to happen, due to the electrostatic repulsion between objects with similar charges.**

Stars are born in clouds consisting of dust, hydrogen and helium. **Gravitational forces** cause regions of the cloud to **contract**, causing **heating**. As the core becomes hotter, the atoms lose their electrons and atomic nuclei collide at high speeds. Eventually a temperature is reached where hydrogen nuclei have enough energy to **fuse** together, resulting in the formation of helium nuclei. This in turn releases **energy**, and the star generates light and other forms of electromagnetic radiation from **nuclear fusion**.

The star enters its **main sequence**. In the main sequence:

● energy is released due to **fusion** of hydrogen nuclei in the core

● outward forces due to the high pressure in the core are balanced by gravitational forces.

Our Sun is a small star which is currently in its main sequence. This will end when there is not enough hydrogen left in the core to generate energy at the rate at which it is being radiated. When this happens:

● the Sun will cool and expand to become a **red giant**

● as the Sun expands the core will contract and become hot enough for the fusion of helium nuclei, resulting in the formation of the nuclei of carbon and oxygen

● the Sun will then contract, losing its outer layers and becoming a very hot, dense body called a **white dwarf**

● as energy is no longer being generated, the colour of a white dwarf changes as it cools and it eventually becomes an invisible **black dwarf**.

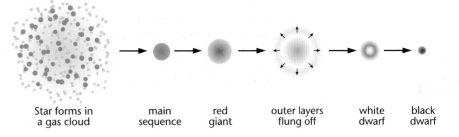

It takes millions of years for a white dwarf to cool and become a black dwarf.

Star forms in a gas cloud    main sequence    red giant    outer layers flung off    white dwarf    black dwarf

**Fig. 4.3** The life cycle of a small star

More massive stars expand to become **red supergiants** after their main sequence. In a red supergiant:

When astronomers notice that the brightness of a star is increasing, they realise that a spectacular event is about to happen when the supernova explodes.

- nuclear fusion in the contracting core results in the formation of nuclei of elements such as magnesium, silicon and iron

- the star is now generating energy again and becomes a **blue supergiant**

- when these nuclear reactions are finished the star cools and contracts again, glowing brightly as its temperature increases and becoming a **supernova**

- the supernova explodes, flinging off the outer layers to form a **dust cloud**

- the core that is left behind is a **neutron star**.

Very dense neutron stars are called **black holes** because they are so dense that even light cannot escape from their gravitational fields. Black holes are detected by their effect on surrounding objects. They pull in gases from nearby stars. These gases reach very high speeds, and emit X-rays which can be detected as evidence for the existence of a black hole.

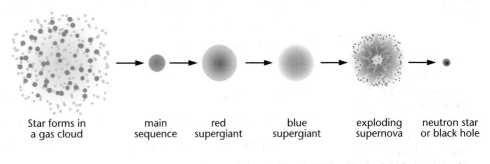

Star forms in a gas cloud    main sequence    red supergiant    blue supergiant    exploding supernova    neutron star or black hole

**Fig. 4.4** The life cycle of a massive star

The existence of heavy elements in the inner planets and the Sun is evidence that our Solar System formed from the gas and dust flung off from the outer layers of an exploding supernova.

# The past and future of the Universe

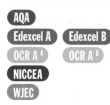

AQA
Edexcel A   Edexcel B
OCR A ᴬ   OCR A ᴮ
NICCEA
WJEC

Evidence about the Universe's past history comes from examining the **spectra** of light given out by stars. When sources of waves such as light and sound move away from an observer, the waves detected by the observer have a longer wavelength than those emitted. This is known as **"red shift"**. When light from stars in other galaxies is analysed the results show that:

- light from almost all galaxies shows **red shift**

- the further away the galaxy, the greater the amount of red shift.

*The apparent change in wavelength of sound waves is readily observed as an aircraft passes overhead or the siren of an emergency vehicle approaches and then recedes.*

This suggests that the Universe is currently expanding. One theory that explains this expansion is the **"Big Bang"**. According to this model:

- the Universe started at a single point with an enormous explosion

- since the dawn of time, the Universe has been expanding and cooling

- the **microwave energy** that fills space is radiation left over from the explosion.

*The age of the Universe is only an estimate due to uncertainties in measuring the speeds of the galaxies and conjecturing how these speeds have changed over millions of years.*

The "Big Bang" model is used to estimate the age of the Universe from measurements of the speeds of the galaxies. Current estimates show that they would have all been in the same place between 15 and 18 billion years ago.

The future of the Universe depends on the speeds at which the galaxies are moving apart and the amount of mass it contains. The three options are:

- there is enough mass for gravitational forces between the galaxies to stop the expansion and cause the Universe to contract, ending in a **"Big Crunch"**

- there is just enough mass to stop the expansion, leaving the Universe in a steady state

- there is insufficient mass to stop the expansion, and the Universe will continue to expand and cool.

**PROGRESS CHECK**

1. What reaction occurs in the core of a star in its main sequence?
2. How does a black hole get its name?
3. What TWO pieces of evidence support the "Big Bang" theory of the origin of the Universe?

1. The fusion of hydrogen nuclei to form the nuclei of helium;   2. It cannot be seen because light cannot escape from it;   3. The red shift of light from other galaxies and the microwave radiation that fills space.

# Sample GCSE question

**1.** There are thousands of artificial satellites in orbit around the Earth.

**(a)** Give TWO uses of artificial satellites. **[2]**

> *Artificial satellites are used for monitoring the weather ✓ and for television transmissions between the UK and the USA ✓.*

*Alternative acceptable answers include navigation, surveillance and for space telescopes.*

**(b)** Some satellites orbit the Earth above the equator with an orbit time of 24 hours. These are called geostationary satellites. Explain why 24 hours is a suitable orbit time for some satellites. **[2]**

> *This is the time it takes for the Earth to rotate on its axis ✓, so the satellite is always above the same point on the Earth's surface ✓.*

*A common error at GCSE is to state that a geostationary satellite moves at the same speed as the Earth – this is not the case.*

**(c)** The graph shows how the orbit time of a satellite depends on its height above the Earth's surface.

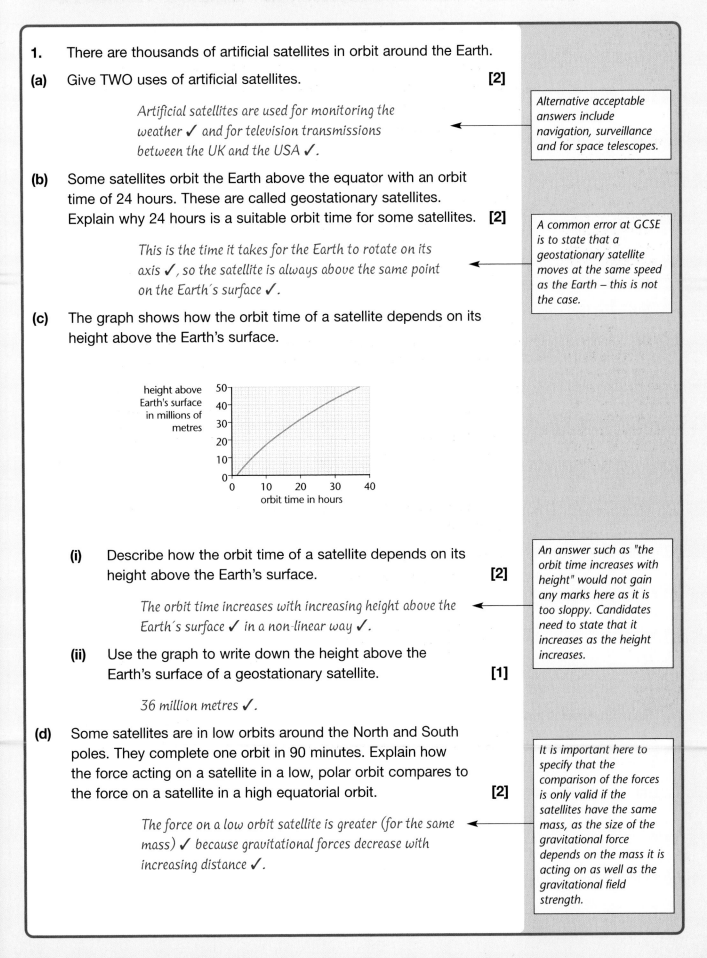

height above Earth's surface in millions of metres

orbit time in hours

**(i)** Describe how the orbit time of a satellite depends on its height above the Earth's surface. **[2]**

> *The orbit time increases with increasing height above the Earth's surface ✓ in a non-linear way ✓.*

*An answer such as "the orbit time increases with height" would not gain any marks here as it is too sloppy. Candidates need to state that it increases as the height increases.*

**(ii)** Use the graph to write down the height above the Earth's surface of a geostationary satellite. **[1]**

> *36 million metres ✓.*

**(d)** Some satellites are in low orbits around the North and South poles. They complete one orbit in 90 minutes. Explain how the force acting on a satellite in a low, polar orbit compares to the force on a satellite in a high equatorial orbit. **[2]**

> *The force on a low orbit satellite is greater (for the same mass) ✓ because gravitational forces decrease with increasing distance ✓.*

*It is important here to specify that the comparison of the forces is only valid if the satellites have the same mass, as the size of the gravitational force depends on the mass it is acting on as well as the gravitational field strength.*

# Exam practice questions

1. The orbit time of a satellite depends on its height above the Earth's surface. A satellite used for television broadcasts orbits the Earth directly above the equator and has an orbit time of 24 hours.

   The diagram shows such an orbit.

   **(a)** Explain why this is called a geostationary orbit. **[2]**

   **(b)** Explain why satellites used for television broadcasts use geostationary orbits. **[2]**

   **(c)** Some communications satellites used for mobile telephones are in elliptical orbits. The diagram shows a satellite in an elliptical orbit.

   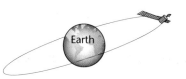

   **(i)** Draw an arrow that shows the gravitational force on the satellite. **[1]**

   **(ii)** Describe how the force on the satellite changes as it approaches the Earth. **[2]**

   **(iii)** At which point on its orbit does the satellite have its greatest speed? **[1]**

2. Stars are formed within clouds consisting mainly of hydrogen, helium and dust.

   **(a)** Describe how a star is formed in such a cloud. **[3]**

   **(b)** Our Sun is a small star that also contains more massive elements. What does the presence of these elements suggest about the origin of the Sun?

   Give the reason for your answer. **[2]**

   **(c)** Our Sun is currently in its main sequence.

   **(i)** What happens to a star in its main sequence? **[2]**

   **(ii)** What is likely to happen to the Sun at the end of its main sequence? **[3]**

# Exam practice questions

**3.**

**(a)** What evidence is there that the galaxies are moving away from each other? **[3]**

**(b)** The Andromeda galaxy is moving towards our galaxy, the Milky Way.
How does light detected from the Andromeda galaxy differ from that detected from other galaxies? **[2]**

**(c)** **(i)** What additional evidence is there to support the "Big Bang" theory? **[1]**

**(ii)** How does this theory picture the evolution of the Universe up to the present time? **[3]**

**(d)** One possible future of the Universe is the "Big Crunch".
What has to happen to cause this? **[3]**

**4.**

**(a)** In 1610 Galileo trained his newly-discovered telescope on Jupiter and discovered four star-like objects.

**(i)** Suggest why no-one had reported seeing these before. **[1]**

After observing their position for several days, he concluded that they were moons of Jupiter.

**(ii)** What could he have observed to lead to this conclusion? **[2]**

**(b)** The table gives some data about the Galilean moons.

| Moon | Orbital distance in m | Orbital period in s | Radius in m | Density in g/cm$^3$ |
|------|----------------------|--------------------|-------------|--------------------|
| Io | $4.3 \times 10^8$ | $1.5 \times 10^5$ | $1.8 \times 10^6$ | 3.6 |
| Europa | $6.7 \times 10^8$ | $3.1 \times 10^5$ | $1.6 \times 10^6$ | 3.0 |
| Ganymede | $1.1 \times 10^9$ | $6.2 \times 10^5$ | $2.6 \times 10^6$ | 1.9 |
| Callisto | $1.9 \times 10^9$ | $1.4 \times 10^6$ | $2.4 \times 10^6$ | 1.8 |

**(i)** What is the relationship between orbital period and orbital distance? **[1]**

**(ii)** What evidence in the table indicates that the two outer moons are icy while the two inner moons are rocky? **[1]**

**(iii)** In what other way are the two outer moons different to the two inner moons? **[1]**

**(iv)** One feature of the Galilean moons is similar to that of the Earth's moon; they each have the same face pointing towards Jupiter at all times.

What is the time period of rotation of Ganymede on its own axis? **[1]**

**The following topics are covered in this section:**

- **Energy transfer and insulation**
- **Work, efficiency and power**
- **Generating and distributing electricity**

## What you should know already

Use words from the list to complete the passage and label the diagram.

You can use each word more than once.

| conduction | convection | conservation | cooler | current | evaporation |
|---|---|---|---|---|---|
| fossil | generated | heat | iron | light | magnetic |
| non-renewable | particles | radiation | relays | renewable | Sun |

Most of the Earth's energy comes from the 1.___Sun___. Resources such as coal, oil and gas, which have stored energy over millions of years are called 2.___fossil___ as they cannot be replenished within the Earth's lifetime. Food, wind and waves are 3.___non-renewable___ resources; they will never run out.

Electricity is 4.___generated___ from both types of resource, but most of our electricity comes from 5.___light___ fuels. Energy from electricity is transferred to movement, heat and 6.___current___ in the appliances we use at home and at work.

Complete the labels to show the energy transfers in the diagrams.

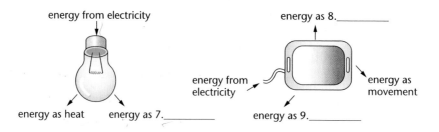

energy from electricity

energy as 8._____

energy from electricity

energy as movement

energy as heat    energy as 7._____    energy as 9._____

There is a constant transfer of energy from hot objects to 10._____ ones. Energy transfer from particle to particle is called 11._____. Energy transfer by the upward and downward movement of fluids is called 12._____. Liquids and other objects containing moisture lose energy by 13._____.

Energy from the Sun travels to the Earth as 14._____. All objects lose and gain energy by this method. It is the only way in which energy is transferred that does not involve the movement of 15._____.

When energy is transferred, there is no gain or loss of energy. This is known as energy 16._____. It does become more spread out, which makes it difficult to recover.

A current passing in a coil of wire has a 17._____ field pattern similar to that of a bar magnet. An electromagnet is made by wrapping a coil of wire around an 18._____ core. The strength of an electromagnet depends on the number of turns of wire and the 19._____. Electromagnets are used to operate switches called 20._____.

# 5.1 Energy transfer and insulation

**After studying this section you should be able to:**

- explain the mechanisms of energy transfer by conduction, convection and radiation
- understand the role of trapped air in insulating houses and people.

## The nature of the surface

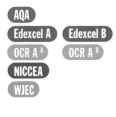

All objects emit and absorb energy in the form of **infra-red radiation**, electromagnetic waves with wavelengths longer than light but shorter than microwaves. The rate at which energy is emitted depends on the temperature and the nature and colour of the surface. The rate at which energy is absorbed depends only on the nature and colour of the surface.

The range of wavelengths emitted depends on the temperature of the object. Hot objects also emit light.

 **KEY POINT** Dark, dull surfaces are good emitters and absorbers of infra-red radiation. Light, shiny surfaces are poor emitters and absorbers of infra-red radiation.

Infra-red radiation is reflected in the same way as light. For this reason:

- marathon runners are often given aluminium foil capes to wear after a race; it reflects back the infra-red radiation emitted by their bodies

- even though it conducts heat, aluminium foil keeps food taken out of an oven hot, as it reduces energy loss by radiation

- in hot countries houses are often painted white to reduce the energy absorbed from the Sun's radiation.

Examples of objects that are painted black to maximise the absorption or emission of radiant energy include:

- the pipes at the rear of a refrigerator or freezer, as the coolant flowing through these pipes is warm and needs to lose the energy absorbed from the inside of the cabinet

Solar panels used to generate electricity are also painted black to maximise the energy absorbed from the Sun's radiation.

- solar heating panels that absorb radiant energy from the Sun and use it to heat water.

## Moving fluids

The movement of a gas or a liquid in a **convection current** is due to parts of the fluid having different **densities**. When a liquid or a gas is heated:

- the particles gain more kinetic energy, causing the fluid to expand

- this results in a decrease in the density of the fluid

- the warmer, less dense fluid rises and is replaced by colder, denser fluid.

The reverse happens when a fluid is cooled. Central heating radiators rely on upwards-driven convection currents to circulate the warm air in a room while refrigerators rely on downwards-driven convection currents to keep the contents cool. These are shown in the diagram, **Fig. 5.1**.

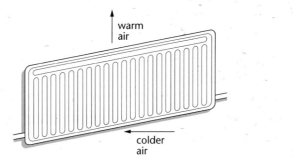

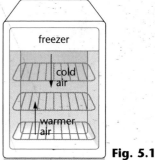

**Fig. 5.1**

## Conduction by particles

All materials allow energy to pass through them by **conduction**. As particles become more energetic, some of this energy is transferred to neighbouring particles as they interact. The particles of a gas are more widespread than those of a solid or a liquid, so interactions between them are less frequent. This is why gases are poor thermal conductors.

Metals are better conductors than non-metals because the free electrons responsible for conduction of electricity in metals also play a role in thermal conduction. These electrons:

- move randomly at high speeds

- travel relatively large distances between collisions with the metal ions

- transfer energy rapidly from hot areas of the metal when they move to cooler areas by diffusion.

# Insulating buildings and bodies

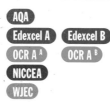

AQA
Edexcel A    Edexcel B
OCR A ^A^    OCR A ^B^
NICCEA
WJEC

Of the three methods of energy transfer described above, the greatest energy loss from warm buildings and bodies occurs through conduction and convection. Energy loss by radiation is more significant when an object is considerably warmer than its surroundings.

Energy is transferred through a glass window, a brick wall and through clothing by conduction. In modern houses the external walls consist of a brick outer wall and a breeze block inner wall. An air gap, called a cavity, separates the two. Energy passes from the warm inside to the cold outside by:

● conduction through the inner wall

● convection through the cavity

● conduction through the outer wall.

> **Energy loss by radiation from the head can be significant. Hair is a good insulator, so people who are bald should always wear a hat in cold weather to reduce this energy loss.**

Energy flow through the wall is reduced by cavity wall insulation. This stops the convection in the cavity by trapping pockets of air. Energy can now only flow through the cavity by conduction. Since air, like all gases, is a poor conductor, the energy flow is much less. Energy flow through the walls of a house is shown in the diagram, **Fig. 5.2**.

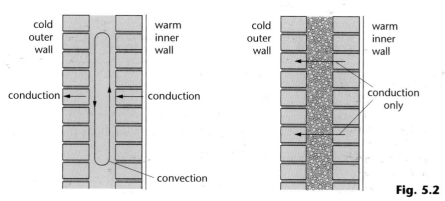

**Fig. 5.2**

Energy flows through a double-glazed window in a similar way. As it is impractical to use another material in the gap between the panes, these should be placed close together. A narrow gap does not allow enough room for convection currents to flow between the glass panes.

Loft insulation is the most cost-effective way of insulating a house. Energy flows through an uninsulated loft in a similar way to an uninsulated external wall:

● energy is conducted through the plasterboard ceiling

● it passes through the air by convection currents

● it is then conducted through the roof tiles.

> **It costs about £100 to install loft insulation in an average house, compared to about £5000 for double-glazed windows.**

To reduce the energy flow through the ceiling, loft insulation in the form of fibreglass is placed above the plasterboard. Fibreglass traps air, which is a good insulator, so less energy is allowed to flow into the airspace to form convection currents in the loft.

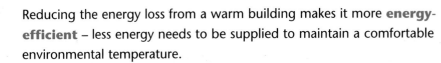

Reducing the energy loss from a warm building makes it more **energy-efficient** – less energy needs to be supplied to maintain a comfortable environmental temperature.

Much of the insulation in a building relies on trapped air. We also use trapped air to insulate our bodies in cold weather. Tight-fitting clothes trap a layer of air next to the skin. In very cold weather the most effective way to reduce the heat loss is to wear more layers of clothing – each extra layer traps another layer of air and reduces the energy transfer by conduction.

**PROGRESS CHECK**

1. Which method of transfer of thermal energy:
   (a) can transfer energy through a vacuum?
   (b) only occurs in fluids?
   (c) can occur in solids, liquids and gases?
2. Explain how cavity wall insulation reduces the energy lost from a warm house.
3. Suggest why several layers of paper provide effective insulation for takeaway food.

1(a) radiation;   (b) convection;   (c) conduction;   2. Air is trapped in pockets. This stops convection currents;   3. A layer of air is trapped between the layers of paper. The trapped air is a poor conductor.

# 5.2 Work, efficiency and power

**LEARNING SUMMARY**

After studying this section you should be able to:

- **describe everyday energy transfers**
- **recall and use the relationships for calculating work, power, kinetic energy and gravitational potential energy**
- **calculate the efficiency of an energy transfer.**

## Work and energy transfer

Everything that happens involves **work** and **energy transfer**. Whenever a force makes something move **work** is being done and **energy** is being **transferred** between objects. The amount of work done and the amount of energy transferred are the same.

 The terms "work" and "energy transfer" have the same meaning.

**KEY POINT**

When a force, *F*, moves an object a distance *x* in its own direction:
**work done = energy transferred = force × distance moved**
$$E = F \times x$$
Work and energy are measured in joules (J) when the force is in N and the distance is in m.

Energy can be transferred and stored in a number of different ways:

- stored energy is **potential energy**; it can be **gravitational** due to position, **elastic** in a stretched spring or **chemical** in a lump of coal or a battery

- the energy stored in the atomic nucleus is **nuclear energy**; this is the energy source in a nuclear power station

- energy due to movement is **kinetic energy**; this includes the energy of a sound wave and the energy of particles in a gas

- the energy of an object due to its temperature is **thermal energy**; the thermal energy of a solid and a liquid comprises both kinetic and potential energy of the particles

- **thermal energy** (heat) is transferred between objects by the processes of conduction, convection, evaporation and radiation

- energy is transferred by an **electric current** from a power supply to the components in a circuit

- energy is transferred between objects by electromagnetic radiation, or **radiant energy**.

## Efficiency of energy transfer

In a coal-fired power station, for every 100 J of energy stored in the coal that is burned, only 40 J is transferred to electricity. **Energy**, like **mass** and **charge**, is a **conserved quantity**. This means that the total amount of energy remains the same, it cannot be created from or turned into a different quantity. In the case of the power station, 60% of the energy input is wasted and ends up as **heat** in the surrounding atmosphere and river.

This is shown in the energy flow diagram, **Fig. 5.3**.

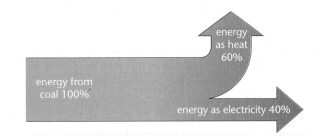

Fig. 5.3

Gas-fired power stations are more **efficient**; they transfer more of the energy from the fuel to electricity. The **efficiency** of an energy transfer is the fraction or percentage of the energy input that is transferred to the desired output.

 **KEY POINT**

$$\text{efficiency} = \frac{\text{useful energy output}}{\text{total energy input}}$$

# Gravitational potential energy and kinetic energy

In the **pumped storage system** shown in the diagram, **Fig. 5.4**, surplus electricity generated at night when the demand is low is used to pump water from a low lake to a high one.

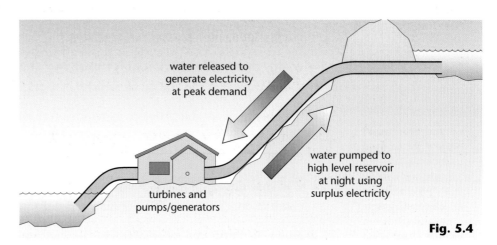

water released to generate electricity at peak demand

water pumped to high level reservoir at night using surplus electricity

turbines and pumps/generators

**Fig. 5.4**

The energy is stored as **gravitational potential energy**.

> **KEY POINT**
> The change in gravitational potential energy when a mass, *m*, moves through a vertical height, *h*, is calculated using the relationship:
> gravitational potential energy = mass × gravitational field strength × height
> GPE = $m \times g \times h$

At times of peak demand, the water is allowed to fall through vertical pipes, transferring energy from **gravitational potential energy** to **kinetic energy** as it does so. The **kinetic energy** of the moving water is transferred to the turbines. The turbines drive the generators in which the kinetic energy is transferred to electrical energy.

A pumped storage system can supply electricity within two minutes after the valves are opened. It takes several hours for a coal-burning power station to become operational.

> **KEY POINT**
> The kinetic energy of a mass, *m*, moving at speed *v*, is calculated using the relationship:
> kinetic energy = $\frac{1}{2}$ × mass × speed$^2$
> KE = $\frac{1}{2} \times m \times v^2$

The **power** output of the generator depends on its efficiency and the power input from the moving water. **Power** is the **rate of energy transfer**, the work done or the energy transferred each second.

At GCSE, candidates often lose marks by confusing the units of energy and power.

> **KEY POINT**
> Power is calculated using the relationship:
> power = $\dfrac{\text{work done or energy transfer}}{\text{time}}$
> $P = \dfrac{E}{t}$
> Power is measured in watts (W) when the energy is in J and the time is in s.

All electrical appliances have a power rating. This is the electrical power input. The output power depends on the **efficiency** of the appliance. Appliances used for heating are much more efficient than those used for lighting and movement.

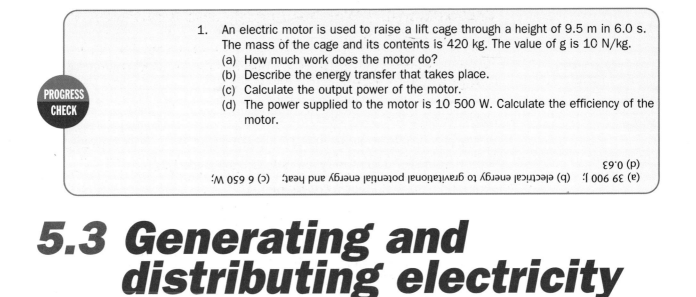

1. An electric motor is used to raise a lift cage through a height of 9.5 m in 6.0 s. The mass of the cage and its contents is 420 kg. The value of g is 10 N/kg.
   (a) How much work does the motor do?
   (b) Describe the energy transfer that takes place.
   (c) Calculate the output power of the motor.
   (d) The power supplied to the motor is 10 500 W. Calculate the efficiency of the motor.

PROGRESS CHECK

(a) 39 900 J;   (b) electrical energy to gravitational potential energy and heat;   (c) 6 650 W;
(d) 0.63

# 5.3 Generating and distributing electricity

After studying this section you should be able to:

LEARNING SUMMARY

● **describe how the magnetic force on a current is used in a motor**
● **explain how electromagnetic induction is involved in generating and distributing electricity**
● **evaluate the use of different energy resources for producing electricity.**

## The motor effect

AQA
Edexcel A    Edexcel B
OCR A A     OCR A B
NICCEA
WJEC

There is a **force** on an **electric current** that passes in a direction at right angles to a **magnetic field**.

The direction of this force:

● is at right angles to both the magnetic field and the current direction

● is reversed if either the current or the magnetic field direction is reversed.

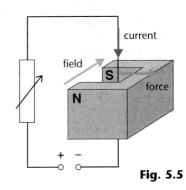

**Fig. 5.5**

The strength of the magnetic field can be increased by using stronger magnets.

The size of the force:

● is increased by increasing the current or the strength of the magnetic field.

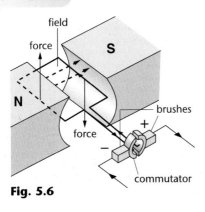

**Fig. 5.6**

The diagram, **Fig 5.6**, shows how this is used to produce rotation in a simple **d.c. motor**. The purpose of the **commutator** is to reverse the direction of the current as the coil passes through the vertical position. This is necessary to keep the loop turning in the same direction.

The motor effect is also used to produce the vibrational movement of a **loudspeaker** cone. A coil of wire within a magnetic field carries an **alternating current**. The force on the coil is reversed, reversing the direction of movement, whenever the current changes direction.

# Electromagnetic induction

AQA
Edexcel A  Edexcel B
OCR A ᴬ  OCR A ᴮ
NICCEA
WJEC

When the **magnetic field** through a coil changes, it causes a **voltage** across the terminals of the coil. This voltage is called an induced voltage and the phenomenon is known as **electromagnetic induction**. A voltage can be induced in a conductor by:

> When answering questions about electromagnetic induction, always emphasise the change in the magnetic field.

- moving the conductor at right angles to a magnetic field
- moving a magnet inside a coil of wire
- switching a nearby electromagnet on or off.

> Moving the coil around a stationary magnet would have a similar effect.

The diagram, **Fig. 5.7**, shows a **voltage** being **induced** as a magnet is moved into a coil of wire. A **current** passes if there is a complete circuit.

The size of the induced voltage can be increased by:

- increasing the number of turns on the coil
- increasing the area of the coil
- moving the magnet faster
- increasing the strength of the magnetic field.

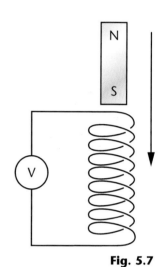

**Fig. 5.7**

The direction of the induced voltage can be reversed by:

- reversing the direction of movement
- reversing the poles of the magnet.

## The generator

> Rotating the coil causes a change in the magnetic field through it.

If a **coil** of wire is rotated within a **magnetic field**, electromagnetic induction causes a **voltage** across the connections to the coil. The diagram, **Fig. 5.8**, shows an **a.c. generator**. The slip rings and brushes enable the induced current to pass out of the coil.

Generators like this are used to provide the electricity supply for motor vehicles.

Power station generators use a rotating **electromagnet** to produce a changing magnetic field. A voltage is induced in thick copper bars around the electro-magnet. The electromagnet rotates at 3000 revolutions each minute to induce a voltage with a frequency of 50 Hz.

> It would be impractical to rotate the copper bars at this speed.

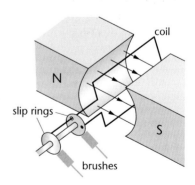

**Fig. 5.8**

Bicycle **dynamos** also use a rotating magnet, see **Fig 5.9**. As a bicycle speeds up the magnet rotates faster, increasing the size of the induced voltage so the lights become brighter. It also increases the **frequency** of the induced voltage, since one cycle is generated for each revolution of the magnet.

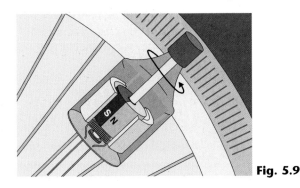

**Fig. 5.9**

## The transformer

A **transformer** changes the size of an **alternating voltage**. The diagram, **Fig. 5.10**, shows the construction of a transformer. It works on the principle of electromagnetic induction:

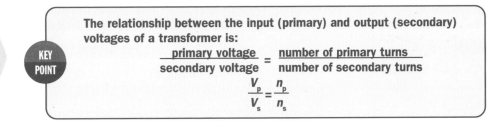

iron core

a.c. source

primary coil          secondary coil

**Fig. 5.10**

- an alternating current in the input or primary coil produces a changing magnetic field

- this changing field is concentrated in the iron core, so that it passes through the output, or secondary coil

- the **changing magnetic field** induces a **voltage** in the secondary coil.

The size of the output voltage depends on the size of the input voltage and the ratio of the numbers of turns on the coils.

> This formula is cumbersome to use. When calculating transformer voltages, it is usually easier to work in ratios.

> **KEY POINT**
>
> The relationship between the input (primary) and output (secondary) voltages of a transformer is:
>
> $$\frac{\text{primary voltage}}{\text{secondary voltage}} = \frac{\text{number of primary turns}}{\text{number of secondary turns}}$$
>
> $$\frac{V_p}{V_s} = \frac{n_p}{n_s}$$

This means that the voltages are in the **same ratio** as the **numbers of turns** on the coils:

- a **step-up** transformer **increases** the voltage; it has more turns on the secondary than on the primary

- a **step-down** transformer **decreases** the voltage; it has fewer turns on the secondary than on the primary.

When a transformer is used to **increase a voltage**, the **current is reduced** by the same factor. Similarly, decreasing the size of a voltage results in an increased current.

## Transmitting power

Whenever a current passes in a wire, the resistance of the wire causes **energy loss** due to heat. The energy losses can be minimised by having very low resistance wires. This leads to conflict between installation costs and running costs. The conflict is resolved by using transformers:

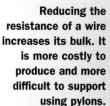
● the rate of energy transfer to heat in transmission wires is proportional to the (current)$^2$

● **low currents** need to be used to minimise energy losses

● this can be achieved by transmitting power at **high voltages**.

A typical power station generator produces electricity at a voltage of 25 000 V. This is stepped up by a transformer before passing into the grid at 400 000 V. The high voltage electricity is stepped down in stages before it reaches consumers such as transport, industry and houses. This process is shown in the diagram, **Fig. 5.11**.

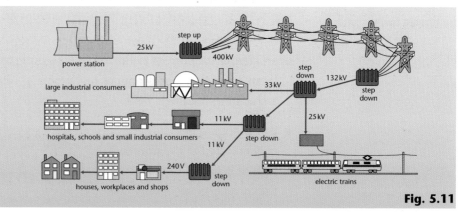

Fig. 5.11

Transformers provide an efficient way of changing the voltage of an alternating current. There is no equivalent way of changing the voltage of a direct current, which is why we use alternating current for mains electricity.

# Energy resources

AQA
Edexcel A   Edexcel B
OCR A$^A$   OCR A$^B$
NICCEA
WJEC

Most of the electricity generated in the UK comes from **fossil fuels** – coal, gas and oil. The amount of electricity produced from nuclear energy is significant, but decreasing. Some electricity is generated from moving water in fast-flowing rivers and tides and a small amount comes from **wind-powered** generators and **geothermal** energy.

In a **coal-fired** power station:

● energy obtained from burning coal is used to turn water into steam at high temperature and pressure

● this steam drives the **turbines** which in turn drive the **generators**

● a significant energy loss occurs when the steam is condensed back into water to be pumped back to the boiler.

In a **gas-fired** power station:

- gas is burned in a combustion chamber
- the hot exhaust gases drive the turbines directly
- the energy remaining in the exhaust gases is used to generate steam to drive a steam turbine
- this process has an overall **efficiency** of 50% – 10% greater than a coal-fired power station.

Although the known reserves of coal are greater than those of gas, new gas fields are constantly being discovered. The building of gas-fired stations to replace older coal-fired ones has helped the UK to reduce its **carbon dioxide** emissions in recent years.

There are no power stations in the UK currently burning only oil, though oil is used to pre-heat the boilers in coal-fired stations.

**Burning gas causes less sulphur dioxide pollution than oil, which in turn causes less than coal.**

The burning of fossil fuels causes **pollution** in the forms of **carbon dioxide**, a greenhouse gas, and **sulphur dioxide**, which contributes to **acid rain**. Sulphur dioxide can be removed from power station exhaust gases by passing them through a slurry of powdered limestone mixed with water. However, this leads to other forms of pollution in mining and transporting the limestone.

In a **nuclear** power station:

- energy released from uranium and plutonium is used to create steam and drive a steam turbine, as in a coal-fired station
- there are **environmental hazards** due to the accidental release of radioactive material
- the **waste** materials can be highly **radioactive** and present long-term disposal and storage problems.

**The high cost of closing down a nuclear power station adds significantly to the cost of the electricity produced.**

The use of renewable energy resources such as **wind** and **moving water** has the advantage of producing no atmospheric pollution:

- fast-flowing rivers and streams are used to drive water turbines in the production of **hydro-electricity**; suitable rivers and streams are only found in Wales and Scotland

**Building a barrier across a tidal estuary can cause silting-up and consequent environmental damage.**

- there is a vast amount of energy available from **tides**; this does not depend on rainfall or wind and is constantly available but attempts to transfer this energy to electricity have proved costly and unreliable
- in some parts of the country the wind blows all the time so **wind-powered** generators can be relied upon; however wind farms have a high installation cost, they are regarded as noisy and unsightly by some people and they take up a large area to produce a relatively small amount of electricity
- underground rocks are constantly being heated by **radioactive decay**; the **geothermal energy** can be extracted by pumping water through them, but there are few sites in the UK where the rocks are hot enough to generate steam capable of driving turbines.

## Capturing energy from the Sun

**Radiant energy** from the Sun can be used to heat water and generate electricity. The use of **solar heating** to provide domestic hot water is very common in southern Europe where the intensity of the Sun's radiation is greater than that in the UK, so it takes less time to recoup the capital outlay in lower fuel bills.

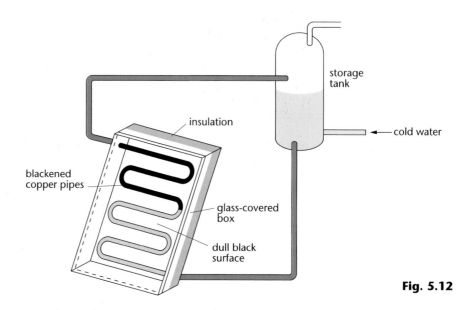

storage tank

insulation

cold water

blackened copper pipes

glass-covered box

dull black surface

**Fig. 5.12**

The diagram, **Fig. 5.12**, shows a solar heated water panel and hot water storage tank:

● **short-wavelength** radiation from the Sun passes through the glass panel and is absorbed by the blackened copper pipes

● energy passes through the copper pipes by **conduction** to heat the water

● **long-wavelength** infra-red radiation emitted by the water pipes does not pass through the glass panel; much of it is reabsorbed by the pipes.

**Photovoltaic cells** can be used to clad the walls and roofs of buildings and generate electricity. At the moment these have a very low **efficiency**, around 20%, and a very high capital cost. They are useful for low-power appliances such as calculators and to power telephone boxes in remote areas, where it is cheaper to install the solar cells than to provide a connection to the mains electricity supply.

**PROGRESS CHECK**

1. What two changes take place to the output of an a.c. generator when it is turned faster?
2. A transformer has an input voltage of 240 V and 1 000 turns on the primary coil. How many turns are needed on the secondary coil for the output voltage to be 12 V?
3. Explain why electricity is transmitted at high voltage and low current.

1. The voltage and frequency both increase;    2. 50;    3. To minimise energy lost as heat in the wires.

# Sample GCSE question

1. Two coils of wire are wound on an iron core.
   One coil is connected, through a switch, to a d.c. supply.
   The second coil is connected to a sensitive ammeter.

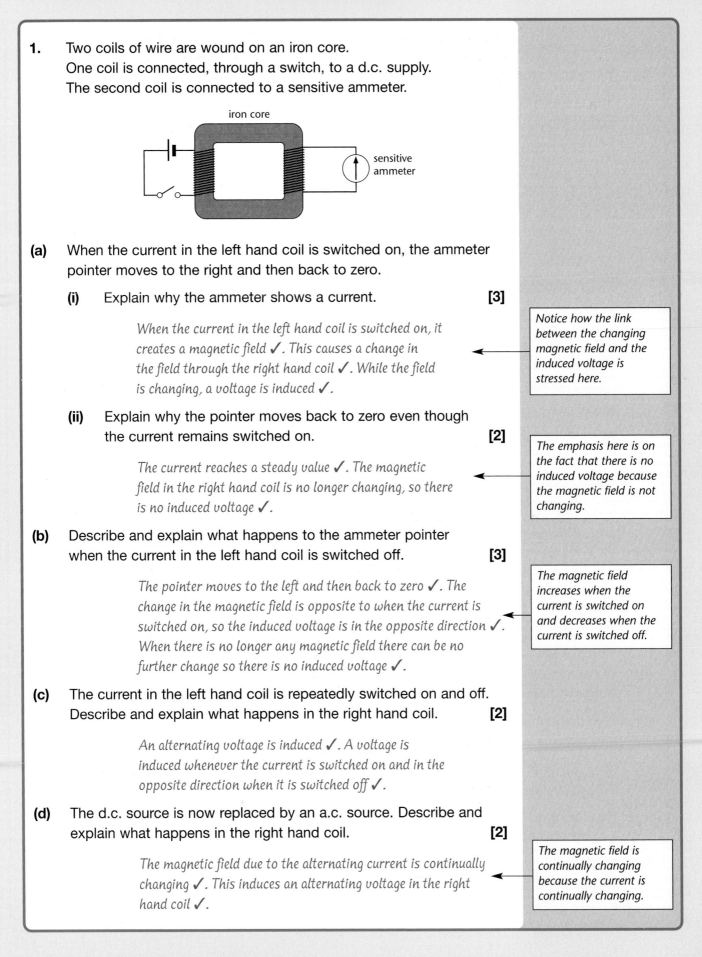

iron core

sensitive ammeter

**(a)** When the current in the left hand coil is switched on, the ammeter pointer moves to the right and then back to zero.

   **(i)** Explain why the ammeter shows a current. **[3]**

   *When the current in the left hand coil is switched on, it creates a magnetic field ✓. This causes a change in the field through the right hand coil ✓. While the field is changing, a voltage is induced ✓.*

   *Notice how the link between the changing magnetic field and the induced voltage is stressed here.*

   **(ii)** Explain why the pointer moves back to zero even though the current remains switched on. **[2]**

   *The current reaches a steady value ✓. The magnetic field in the right hand coil is no longer changing, so there is no induced voltage ✓.*

   *The emphasis here is on the fact that there is no induced voltage because the magnetic field is not changing.*

**(b)** Describe and explain what happens to the ammeter pointer when the current in the left hand coil is switched off. **[3]**

   *The pointer moves to the left and then back to zero ✓. The change in the magnetic field is opposite to when the current is switched on, so the induced voltage is in the opposite direction ✓. When there is no longer any magnetic field there can be no further change so there is no induced voltage ✓.*

   *The magnetic field increases when the current is switched on and decreases when the current is switched off.*

**(c)** The current in the left hand coil is repeatedly switched on and off. Describe and explain what happens in the right hand coil. **[2]**

   *An alternating voltage is induced ✓. A voltage is induced whenever the current is switched on and in the opposite direction when it is switched off ✓.*

**(d)** The d.c. source is now replaced by an a.c. source. Describe and explain what happens in the right hand coil. **[2]**

   *The magnetic field due to the alternating current is continually changing ✓. This induces an alternating voltage in the right hand coil ✓.*

   *The magnetic field is continually changing because the current is continually changing.*

# Exam practice questions

1.  A voltage is induced in a conductor when the magnetic field through it changes.

(a)  What is the name of this effect?                                               [1]

(b)  The diagram shows a coil of wire next to a magnet. A voltmeter is connected to the coil of wire.

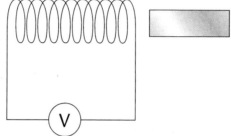

    (i)    Describe TWO ways of inducing a voltage in the coil of wire.          [2]

    (ii)   State THREE factors that affect the size of the induced voltage.       [3]

    (iii)  Write down TWO ways of changing the direction of the induced voltage.   [2]

(c)  A transformer consists of two coils of wire on an iron core. An alternating voltage applied to the input (primary) coil causes an alternating voltage in the output (secondary) coil.

    (i)    A 3 V battery is connected to the input coil.
       Explain why there is no voltage across the output coil.                [2]

    (ii)   A 3 V a.c. source is connected to the input coil.
       Explain why there is a voltage across the output coil.                 [2]

    (iii)  The input coil has 200 turns.
       When the 3 V a.c. source is connected to this coil, the output voltage is
       12 V. Calculate the number of turns on the output coil.               [2]


2.  A power station generator consists of an electromagnet that rotates inside three sets of copper conductors. The current in each conductor is 7500 A.

(a)  Water flows through channels inside the conductors.
    Suggest why this is necessary.                                         [2]

(b)  Electricity is transmitted along the national grid using a combination of overhead and underground cables.

    (i)    Suggest TWO reasons why overhead conductors are used in preference
       to underground conductors outside towns and cities.                   [2]

    (ii)   Suggest TWO reasons why underground conductors are used in
       preference to overhead conductors in towns and cities.               [2]

(c)  Explain why renewable sources of energy are used to produce only a small proportion of the electricity generated in the United Kingdom.   [3]

# Exam practice questions

**3. (a)** Explain how the construction of a step-up transformer differs from that of a
step-down transformer. **[3]**

**(b)** Describe how transformers are used in the distribution of electricity. **[2]**

**(c)** Explain why it is necessary to distribute electricity at high voltage. **[2]**

**(d)** Explain why mains electricity is alternating current rather than direct current. **[2]**

**4.** Electricity can be generated from renewable and non-renewable energy resources.

**(a)** Name TWO renewable energy resources. **[2]**

**(b)** Name TWO non-renewable energy resources. **[2]**

**(c)** Explain how gas-fired power stations cause less damage to the environment
than coal-fired power stations. **[3]**

**(d)** Suggest why it is a social advantage to burn British coal in coal-fired
power stations. **[2]**

**(e) (i)** Give ONE advantage and TWO disadvantages of generating electricity
from wind power. **[3]**

**(ii)** Suggest why very little electricity is generated in the UK from moving
water. **[3]**

**(iii)** France has limited fossil fuel reserves, mostly coal. Most of its electricity is
generated from nuclear power. Outline the advantages and disadvantages
of generating electricity from nuclear power. **[4]**

**5.** The diagram shows a d.c. motor.

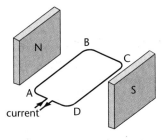

**(a)** Draw an arrow on the diagram that shows the direction of the magnetic field. **[1]**

**(b) (i)** Explain why, in the position shown, the sides AB and CD of the coil
experience a force. **[1]**

**(ii)** Explain why there is no force on the sides BC and AD. **[1]**

**(c) (i)** Explain why the forces on the coil cause it to rotate. **[2]**

**(ii)** State TWO ways in which the direction of rotation can be reversed. **[2]**

**(iii)** State TWO ways in which the speed of rotation can be increased. **[2]**

# Radioactivity

**The following topics are covered in this section:**

● **Ionising radiations** ● **Using radiation**

# 6.1 Ionising radiations

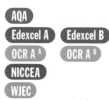

**LEARNING SUMMARY**

*After studying this section you should be able to:*

● **distinguish between the three main radioactive emissions in terms of their penetration and ionisation**
● **describe the effects of radiation on the body**
● **explain the effect on the nucleus when it decays by emitting radiation.**

## Radiation from the nucleus

AQA
Edexcel A     Edexcel B
OCR A ᴬ      OCR A ᴮ
NICCEA
WJEC

Beta radiation causes more ionisation in the materials it passes through than gamma radiation, but less than alpha radiation.

If a material is **radioactive** its atomic nuclei are **unstable**; when they change to a more stable form they emit **radiation**. Some materials are naturally radioactive, others become radioactive when the nucleus absorbs neutrons, as happens in a nuclear reactor. The three main types of nuclear radiation are:

● **alpha** ($\alpha$) – this is intensely ionising but has a very low penetration, being absorbed by a few centimetres of air or a sheet of paper

● **beta** ($\beta$) – this is strongly ionising and more penetrative than alpha radiation. It is partly absorbed by thick paper or card and completely absorbed by a few millimetre thickness of aluminium or other metal

● **gamma** ($\gamma$) – this is only weakly ionising but very penetrative; its intensity is reduced by thick lead or concrete.

**KEY POINT**

an alpha particle consists of two protons and two neutrons
a beta particle is a fast-moving electron
gamma radiation is a high-frequency, short-wavelength electromagnetic wave

All three types of radiation can be detected by a **Geiger-Müller** tube and by **photographic film**, which blackens by exposure to radiation in the same way as it does when it is exposed to light.

### Radiation all around us

Everything we eat depends on plants, so radioactive carbon absorbed by plants is present in all our food.

We are constantly being bombarded by radiation from our surroundings, called the **background radiation**. A radioactive form of carbon, **carbon-14**, is created in the atmosphere and absorbed by plants, so radioactivity is present throughout all food chains. Background radiation is due to:

● radioactivity in all plants and animals

● radiation from the Sun and space

- radiation from buildings

- radiation from the ground

- radiation from hospitals and industrial users of radioactive materials

- radiation from waste materials and "leaks" from nuclear power stations.

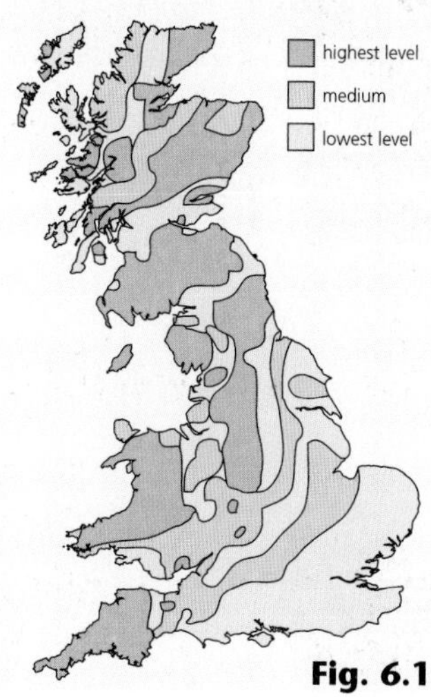

Fig. 6.1

**Granite contains radium, which decays by the emission of an alpha particle to form radon, a radioactive gas which seeps out of the rock.**

Granite is a radioactive rock that emits a radioactive gas called **radon**. This gas accumulates in buildings and is a particular hazard when breathed in, as it emits **alpha** radiation. The occurrence of granite is one reason why the level of background radiation is higher in some parts of the country than in others. The variation in the levels of background radiation is shown in the diagram, **Fig. 6.1**.

Radiation can affect the body in a number of ways:

- it can destroy **cells** and **tissue**

- it can also change the DNA, causing **mutations** and affecting future generations if the sex cells are affected

- radiation can cause **skin burns** and **cancer**.

**Alpha** radiation is particularly hazardous when it enters the body because of its ability to damage cells. It is less of a threat when outside the body because of its low penetration.

**Beta** radiation is a hazard both inside the body and from the outside. Its penetration allows it to pass through the skin and be absorbed by body tissue, where its **ionising ability** enables it to cause damage to cells and tissue.

**In radiotherapy, gamma emitters target specific areas of the body. Cancer cells are affected more than normal cells because they reproduce more rapidly.**

Although it is the most penetrative radioactive emission, **gamma** radiation is less hazardous than beta radiation because it is less likely to be absorbed by body tissue. When it is absorbed, it can destroy cells and is used in radiotherapy treatment to kill cancer cells.

To reduce the hazards from radioactive materials:

- people should be **shielded** from radioactive sources by a suitable absorber

- there should be **as large a distance** as possible between a person and a radioactive source

- the time of exposure to radiation should be as **short** as possible

- people who work with radioactive materials wear a badge containing photographic film that measures the amount of exposure to radiation.

# The effect on the nucleus

AQA
Edexcel A   Edexcel B
NICCEA
WJEC

The scattering of alpha particles by thin gold foil shows that the atom is mainly empty space, with a **large concentration** of **mass** and **positive charge** in a tiny volume. This is the atomic **nucleus**, which is surrounded by orbiting electrons. The nucleus contains two types of particle:

> Alpha particle scattering experiments were first carried out by Geiger and Marsden under the guidance of Lord Rutherford.

- **neutrons** which have no charge

- **protons** which have the same mass as neutrons and carry a single **positive** charge.

The orbiting **electrons** have very little mass and each carries a single **negative** charge. A neutral atom contains equal numbers of protons and electrons.

> **KEY POINT**
> the atomic number or proton number (**Z**) is the number of protons in the nucleus
> the mass number or nucleon number (**A**) is the total number of (protons + neutrons) in the nucleus.

The structure of a nucleus is represented in symbol form as $^A_Z$El, for example $^{24}_{12}$Mg represents an atom of magnesium that has twelve protons and twelve neutrons.

The number of protons in the nucleus determines the element, but not all atoms of the same element are identical. Some elements exist in different forms called **isotopes**.

> Hydrogen has two isotopes, called deuterium and tritium. Unlike hydrogen, the nuclei contain neutrons; one in the case of deuterium and two in the case of tritium.

> **KEY POINT**
> Isotopes of an element have the same number of protons but different numbers of neutrons in the nucleus.

The isotopes of an element all have the same chemical properties but vary in the physical properties of density and nuclear stability.

When a radioactive isotope decays by alpha or beta emission, it results in the formation of the nucleus of a different element:

- **alpha** emission causes the loss of **two protons** and **two neutrons**, so the nucleus formed has an atomic number which has decreased by two and a mass number which has decreased by four

- in **beta** emission a **neutron** decays to a **proton** and an **electron**; the atomic number increases by one and the mass number is unaffected

> In beta decay the electron is ejected from the nucleus, leaving it with one more proton and one less neutron.

- **gamma** emission does not affect the mass or atomic number; it is emitted to reduce the excess energy of the nucleus, often following alpha or beta emission.

> Check the balance of these equations. The upper numbers represent mass and the lower numbers represent charge.

The equations that represent radioactive decay are balanced in terms of charge (represented by $Z$, the number of protons) and mass (represented by $A$, the number of nucleons). The decay of carbon-14 by beta-emission and radon-220 by alpha-emission are represented by the equations:

$$^{14}_{6}\text{C} \rightarrow {}^{14}_{7}\text{N} + {}^{0}_{-1}\text{e}$$

$$^{220}_{86}\text{Rn} \rightarrow {}^{216}_{84}\text{Po} + {}^{4}_{2}\propto$$

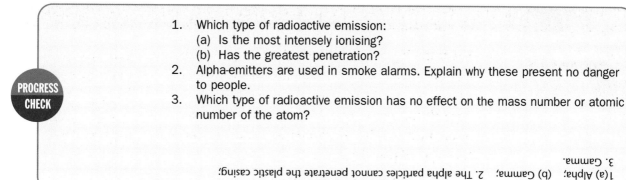
# 6.2 Using radiation

 **LEARNING SUMMARY**

*After studying this section you should be able to:*

● *explain how the decay of a radioactive isotope changes with time*
● *recall and use the term half-life*
● *describe how radioactive isotopes are used to date rocks and other objects.*

## Radioactive decay and half-life

**AQA**
**Edexcel A**　**Edexcel B**
**OCR A A**　**OCR A B**
**NICCEA**
**WJEC**

**Radioactive decay** is a **random** process; the decay of any particular nucleus is unpredictable and, unlike chemical reactions, radioactive decay is not affected by physical conditions such as temperature. The rate at which an **isotope** decays depends on:

● the number of **undecayed nuclei** present in the sample; on average, doubling the number of undecayed nuclei should double the rate of decay

● the **stability** of the radioactive isotope; some isotopes decay much more rapidly than others.

 **KEY POINT**　The rate of decay is the number of nuclei that decay each second. It is measured in **becquerel (Bq)** where 1 Bq = 1 decay/s

 The rate of decay is proportional to the number of undecayed nuclei present in the sample.

As a sample of a radioactive isotope decays, the number of undecayed nuclei decreases, and so the **rate of decay** also decreases. The graph, **Fig. 6.2** shows a typical decay curve.

A graph of rate of decay against time would have the same shape as this, the only difference being the figures on the y-axis.

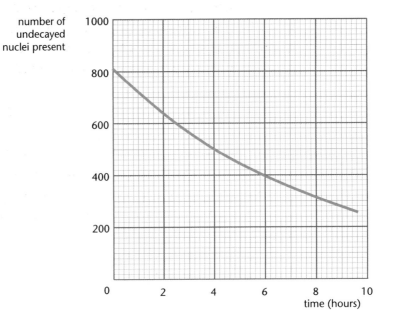

Fig. 6.2

This graph shows that the time it takes for the number of undecayed nuclei to **halve** from 800 to 400 is the same as the time it takes for the number of undecayed nuclei to **halve** from 600 to 300 or between any other two corresponding values. The actual time this takes varies from isotope to isotope but for any one isotope it has a fixed value.

> **KEY POINT** The half-life of a radioactive isotope is the average time it takes for the number of undecayed nuclei to halve.

Because it is difficult to measure the number of undecayed nuclei in a sample, half-life is often measured as the time it takes for the rate of decay to halve.

So, after one half-life, half of the original undecayed nuclei would be expected to remain with one quarter being left after two half-lives have passed.

> **KEY POINT** After $n$ half-lives have elapsed, $\frac{1}{2^n}$ of the original undecayed nuclei present in a sample remain. The rest have changed into nuclei of a different element.

## Radioactive dating

For a more reliable measurement, a mass spectrometer is used to compare the amounts of carbon-12 and carbon-14.

All **living things** are continually absorbing **radioactive carbon-14** from the air or their food. The concentration of carbon-14 in a plant or animal stays at a constant level until it dies. After that it decreases as the carbon-14 decays with a **half-life** of 5730 years and no fresh carbon-14 is absorbed. Archaeological specimens of once-living material such as wood can be dated by measuring the activity of the radioactive carbon remaining.

When **igneous rocks** are formed from **magma**, they contain uranium-238, which decays to form lead-206. The half-life of this decay is 4500 million years. The age of a rock can be dated by comparing the amounts of lead-206 and uranium-238 that it contains. For the oldest rocks found, these isotopes exist in approximately equal amounts, putting the age of the Earth at about 4500 million years.

A similar technique is used for rocks formed containing potassium-40. This has a half-life of 1300 million years; it decays to form argon which can become trapped in the rock. This method can be unreliable because there is no certainty that some of the argon has not seeped out of the rock.

## Nuclear power

**Energy** is released when fission of large nuclei takes place. This is a process in which they are broken up into a number of smaller particles. The fission of uranium-235 is shown in the diagram, **Fig. 14.3.**

> This is a different reaction to that which takes place in stars where energy is released when small nuclei fuse together.

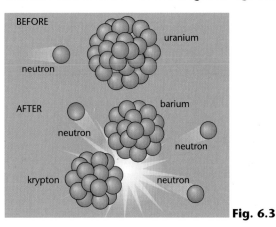

**Fig. 6.3**

In this process:

- a uranium-235 nucleus absorbs a **neutron**, making it **unstable**
- it splits into two smaller nuclei, which are also unstable, and two or three spare neutrons
- the fission products have a lot of **kinetic energy**, which is removed by a coolant
- the coolant generates steam that turns a turbine
- the **spare neutrons** can cause **further fission** of other nuclei.

> In the core of a nuclear reactor, control rods absorb the spare neutrons to control the rate of the reaction.

If each spare neutron were allowed to cause another fission, the result would be a chain reaction which would be out of control. To maintain the reaction at a steady rate, on average just one of the neutrons released by each fission is allowed to go on and cause a further fission.

Nuclear power has one major disadvantage – how to get rid of the waste materials. These are in three categories:

- low-level waste such as laboratory clothing and packaging materials; these are buried either underground or at sea
- intermediate-level waste such as the casing used for nuclear fuel and reactor parts that have been replaced; these are kept in stores with thick concrete walls or buried in deep trenches with concrete linings
- high-level waste such as spent fuel rods; these present a long-term disposal problem since they remain significantly radioactive for thousands of years; much of this waste is in temporary storage in tanks of water until the problem of what to do with it can be solved.

# Some other uses of radioactivity

AQA

Edexcel A

OCR A ᴬ    OCR A ᴮ

NICCEA

WJEC

Radioactive isotopes are also used as **tracers** in medicine and to control the thickness of sheet materials.

When used as a **tracer**:

- the **half-life** of the isotope used should be **long** enough for the tracer still to be radioactive when it has reached its target

- the **half-life** should be **short** enough so that the patient does not remain radioactive for a long period of time, causing unnecessary risk

- the isotope should emit **gamma** radiation **only**; this can be detected outside the body whereas alpha and beta radiations would be strongly absorbed, causing cell damage.

The diagram, **Fig. 6.4**, shows how a radioactive isotope is used to control the thickness of sheet materials.

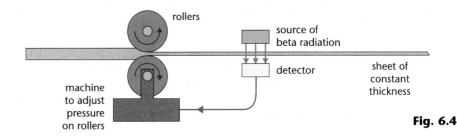

**Fig. 6.4**

> For many sheet materials a beta source is the most suitable.

In this application:

- the radiation from the isotope used must be **partially absorbed** by the material

- if the thickness of the sheet increases, less radioactivity is detected; this information is fed back and the pressure on the rollers is increased

- the isotope should have a **long half-life** to avoid the need for constant recalibration.

---

**PROGRESS CHECK**

1. The rate of decay of a sample of a radioactive isotope depends on two factors. What are these factors?
2. A radioactive isotope with a half-life of 6 hours is used as a tracer in medicine. What fraction of the original nuclei remain after one day?
3. Why is a beta-emitter the most suitable for controlling the thickness of paper?

would be absorbed by the paper.
3. Alpha radiation would be totally absorbed by the paper. No gamma radiation
2. 1/16;
1. The number of undecayed nuclei and the particular isotope in the sample;

# Sample GCSE question

**1.** $^{12}_{6}$C and $^{14}_{6}$C are both isotopes of carbon.

**(a)** **(i)** Write down one similarity about the nucleus of each isotope. **[1]**

*They have the same number of protons ✓.*

**(ii)** Write down one difference in the nucleus of these isotopes. **[1]**

*They have different numbers of neutrons ✓.*

**(b)** $^{14}_{6}$C is radioactive. It decays by emitting a beta particle.

**(i)** Describe a beta particle. **[1]**

*A beta particle is a fast-moving electron ✓.*

**(ii)** Which part of the atom emits the beta particle? **[1]**

*The nucleus ✓.* ⟵

*A common error at GCSE is to state that the beta particle comes from the electrons that orbit the nucleus, since the nucleus does not contain any electrons.*
*All nuclear radiation is emitted from the nucleus; in this case a neutron decays to an electron and a proton.*

**(c)** $^{14}_{6}$C is present in all living materials and in all materials that have been alive. It decays with a half-life of 5730 years. **[2]**

**(i)** Explain the meaning of the term *half-life*.

*Half-life is the average time ✓ for the number of undecayed nuclei to halve ✓.*

**(ii)** The activity of a sample of wood from a freshly-cut tree is measured to be 80 Bq. Estimate the activity of the sample after two half-lives have elapsed. **[1]**

*20 Bq ✓.* ⟵

*A common misunderstanding is that after two half-lives all the nuclei have decayed. This is not the case; on average one half of one half, ie one quarter, of the original nuclei are undecayed after two half-lives.*

**(iii)** The age of old wood can be estimated by measuring its radioactivity. Explain why this method cannot be used to work out the age of a piece of furniture made in the nineteenth-century. **[2]**

*One to two hundred years is a very short time compared to the half-life ✓. The rate of decay would not show any significant change ✓.*

**(iv)** Explain why radiocarbon dating cannot be used to estimate the age of a rock. **[2]**

*Radiocarbon dating works by measuring the decay of carbon-14, which is found in all living things ✓. Rock has never lived, so it does not contain any carbon-14 ✓.*

# Exam practice questions

**1. (a)** The three main types of radioactive emission are called alpha, beta and gamma.

Alpha particles are positively charged.
They consist of two protons and two neutrons.
They are absorbed by thin paper.

Write similar descriptions of beta particles and gamma radiation. **[6]**

**(b)** A source of gamma radiation is pointed at a Geiger-Müller tube connected to a ratemeter.

Sheets of lead of different thicknesses are placed between the source and the Geiger-Müller tube.

The count rate is measured for each lead sheet. The results are shown in the table.

| thickness of lead sheet (mm) | 0 | 5 | 10 | 15 | 20 | 25 |
|---|---|---|---|---|---|---|
| count rate (Bq) | 76 | 57 | 43 | 32 | 24 | 18 |

**(i)** Use a grid to draw a graph of count rate against thickness. **[4]**

**(ii)** What thickness of lead is needed to halve the intensity of the gamma radiation? Explain how you obtain your answer. **[2]**

**(iii)** A similar source emits gamma rays at the rate of 120/s. It is to be transported in a lead container. The emission of gamma rays from the container must not exceed 15/s.

What minimum thickness of container is required? **[3]**

**2.** Technetium-99 is a radioactive isotope used as a tracer in medicine. It decays by emitting gamma radiation only with a half-life of six hours. It is injected into the bloodstream and detected using a camera placed outside the body.

**(a)** Explain the meaning of the terms isotope and half-life. **[4]**

**(b) (i)** Why is it important that the isotope used for this purpose emits gamma radiation? **[2]**

**(ii)** Why is it desirable that the isotope used does not emit alpha or beta radiation? **[2]**

**(c)** Explain why 6 hours is a suitable half-life for an isotope used as a medical tracer.

**3.** A living tree is radioactive. After it dies, the decay rate of carbon-14 in the tree decreases.

**(a) (i)** Suggest why the decay rate does not decrease when the tree is alive. **[1]**

**(ii)** Explain why the decay rate decreases when the tree has died. **[2]**

# Exam practice questions

**(b)** The graph shows how the decay of the carbon-14 in a 0.5 kg sample of wood from the tree changes after the tree has died.

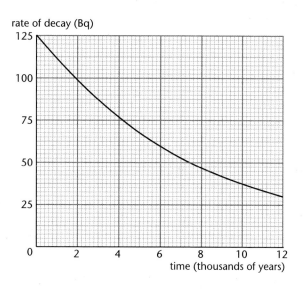

rate of decay (Bq)

time (thousands of years)

**(i)** Estimate the decay rate of a 2 kg sample of wood from a tree immediately after it has died. **[1]**

**(ii)** Use the graph to calculate the half-life of carbon-14.
Show how you obtain your answer. **[2]**

**(iii)** A 0.10 kg sample of wood from an archaeological excavation is found to have a decay rate of 12.0 Bq.

Use the graph to estimate the age of the wood. **[2]**

**4.** Sodium-24 is a radioactive form of sodium that emits gamma radiation. It has a half-life of 15 hours.

In the form of sodium chloride, it is used to detect leaks from underground water pipes.

**(a)** Suggest two reasons why sodium-24 is a suitable isotope to use for this purpose. **[2]**

**(b)** Suggest how it should be used. **[2]**

**(c)** **(i)** What could be used to detect radiation from the water? **[1]**

**(ii)** How would a person operating the detector be able to tell where water was leaking from the pipe? Give the reason for this. **[2]**

**(d)** The water is safe to drink when the radioactivity is one eighth of its initial value. What minimum time should elapse before anyone drinks the water? **[2]**

# Extension material

Check in the table on pages 4 and 5 to see which sections you need to study.

| Topic | Section | Studied in class | Revised | Practice questions |
|---|---|---|---|---|
| 7.1 Electronic control | Electronic systems | | | |
| | A complete system | | | |
| | The timing circuit | | | |
| 7.2 Combining resistors | Resistors in series and in parallel | | | |
| 8.1 Refraction and lenses | Refractive index | | | |
| | Finding the image | | | |
| | Sight | | | |
| 8.2 Resonance | Natural vibrations | | | |
| | Vibrating strings | | | |
| 8.3 Wave interference | What is interference? | | | |
| 8.4 Communicating with waves | Systems | | | |
| | Storing and reading information | | | |
| | Using radio waves | | | |
| | Producing and detecting a radio wave | | | |
| | Using satellites | | | |
| 8.5 Colour | Primary and secondary colours | | | |
| 9.1 Projectiles and momentum | Equations of motion | | | |
| | Adding physical quantities | | | |
| | Types of collision | | | |
| 9.2 Turning in a circle | Stability | | | |
| | Going round in circles | | | |
| 9.3 Some effects of forces | Using springs | | | |
| | Density | | | |
| 10.1 Atoms and nuclei | Farewell to the plum pudding | | | |
| | Nuclear stability | | | |
| 10.2 Electron beams | Producing a beam | | | |
| | Deflecting the beam | | | |
| | Using an oscilloscope | | | |
| 10.3 Particles in motion | Gas pressure | | | |
| | Pressure and temperature | | | |
| | Changing the temperature | | | |
| | | | | |

# 7.1 *Electronic control*

**LEARNING SUMMARY**

*After studying this section you should be able to:*

● *identify the different parts of an electronic system*
● *describe how logic gates are used as processors*
● *explain how a potential divider circuit is used to provide the input to a logic gate.*

## Electronic systems

An electronic system can react to changes in environmental conditions such as pressure, illumination and temperature. There are three components to an electronic system:

● **input sensors** provide information in the form of a voltage that can have different values

● **processors** receive this information and produce an output which depends on the inputs

● **output devices** are operated by the output voltage from the processor.

> In a calculator the keypad provides the input to the processor, and the display is the output device.

Input devices include:

● **switches** – these respond to changes in pressure, magnetic fields or tilting

● **moisture detectors** – these have a resistance that depends on the amount of water present

> In this chapter, the term thermistor refers to one with a negative temperature coefficient – its resistance decreases with increasing temperature.

● **light-dependent resistors (LDRs)** – these have a resistance that depends on the illumination

● **thermistors** – these have a resistance that depends on the temperature.

Output devices include **lamps** and **light-emitting diodes (LEDs)**, but in many applications a relay is used to switch a high-current or high-voltage device such as a heater or a motor.

### Logic gates

> Voltage is measured relative to the negative terminal of the battery or power supply.

Many electronic systems use processors based on **logic gates**. The signals in such a system are **digital** as they can only have one of two values, on (high) or off (low). Most logic gates work from a potential difference (voltage) of 5 V, so an input or output voltage of 0 V is on or low and one of +5 V is on or high. The way in which the output signal depends on the inputs is shown in a **truth table**. In a truth table a 0 represents off and a 1 represents on.

The diagram, **Fig. 7.1**, shows the circuit symbols for NOT, OR and AND logic gates, together with their truth tables.

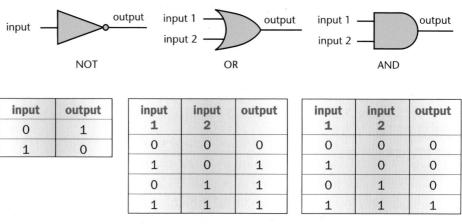

| input | output |
|-------|--------|
| 0 | 1 |
| 1 | 0 |

| input 1 | input 2 | output |
|---------|---------|--------|
| 0 | 0 | 0 |
| 1 | 0 | 1 |
| 0 | 1 | 1 |
| 1 | 1 | 1 |

| input 1 | input 2 | output |
|---------|---------|--------|
| 0 | 0 | 0 |
| 1 | 0 | 0 |
| 0 | 1 | 0 |
| 1 | 1 | 1 |

**Fig. 7.1**

**A NOT gate is also referred to as an invertor.**

The name of each gate describes how the output depends on the inputs:

- NOT – the output is 1 if the input is 0 or NOT 1; this gate inverts an input signal

- OR – the output is 1 if either input 1 OR input 2 is 1 OR both are 1

- AND – the output is 1 if input 1 AND input 2 are both 1.

## Combining logic gates

Outputs from logic gates can be used as inputs to other gates. Combinations of logic gates are used to perform a variety of functions. The diagram, **Fig. 7.2**, shows how NAND and NOR gates can be made from combinations of NOT, AND and OR gates. The truth tables for these gates are also shown.

**NAND means NOT AND and NOR means NOT OR.**

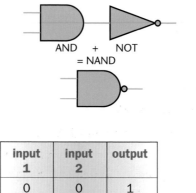

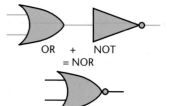

| input 1 | input 2 | output |
|---------|---------|--------|
| 0 | 0 | 1 |
| 1 | 0 | 1 |
| 0 | 1 | 1 |
| 1 | 1 | 0 |

| input 1 | input 2 | output |
|---------|---------|--------|
| 0 | 0 | 1 |
| 1 | 0 | 0 |
| 0 | 1 | 0 |
| 1 | 1 | 0 |

**Fig. 7.2**

The outputs of these gates are the inverse of those of AND and OR.

## Controlling the input

Switches can be used to apply a voltage (0 V or +5 V) directly to the input of a logic gate. Other input sensors are used in **potential divider** circuits to give a voltage that varies with environmental conditions. The diagram, **Fig. 7.3**, shows a potential divider circuit that uses a fixed resistor and a variable resistor.

> Depending on the values of the resistors, the output voltage of a potential divider can range from 0 V to the full supply voltage.

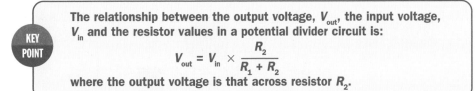

**Fig. 7.3**

In this potential divider circuit:

- the input voltage is equal to the sum of the voltages across the resistors

- the voltage is divided in the ratio of the resistances, so the larger value resistor has the greater voltage across it

- decreasing the resistance of the variable resistor causes the output voltage to rise.

> **KEY POINT**
>
> The relationship between the output voltage, $V_{out}$, the input voltage, $V_{in}$ and the resistor values in a potential divider circuit is:
> $$V_{out} = V_{in} \times \frac{R_2}{R_1 + R_2}$$
> where the output voltage is that across resistor $R_2$.

The diagram, **Fig. 7.4**, shows potential divider circuits that use a thermistor and an LDR to provide the input signals to logic gates.

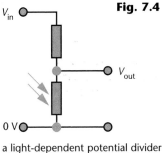

**Fig. 7.4**

> The LDR symbol is sometimes drawn with a circle around the rectangle.

a temperature-dependent potential divider        a light-dependent potential divider

> The threshold voltage is the minimum voltage that the logic gate recognises as being 1. It is usually about half the supply voltage.

In the temperature-dependent potential divider:

- reducing the temperature increases the resistance of the thermistor
- this causes the voltage across the thermistor to rise
- the signal changes from 0 to 1 when the voltage crosses the threshold voltage
- the temperature at which this happens is determined by the value of the fixed resistor; if a variable resistor is used instead this can be adjusted easily.

A similar effect occurs when the illumination of the LDR falls in the light-dependent potential divider.

With the fixed (or variable) resistors and the thermistor or LDR in the **same** positions, the output voltage can be made to fall as the temperature or illumination increases.

## Using the output

An LED can be used to show whether the output of a logic gate is 0 or 1. The output voltage is too high for an LED, so a series resistor is used to limit the current. This is shown in the diagram, **Fig. 7.5**.

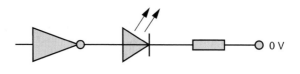

Fig. 7.5

Logic gates can only supply a small current so they can only be used to drive low-power devices. Using the logic gate to switch a relay enables the switching of higher-power appliances which can operate from a separate, independent voltage supply such as the mains.

The diagram, **Fig. 7.6**, shows a logic gate being used to switch a mains-operated lamp. The NO on the relay symbol means that the switch contacts are normally open (switched off). They close, switching on the circuit, when current passes in the relay coil.

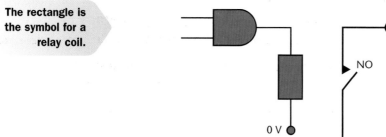

Fig. 7.6

## A complete system

AQA
OCR A ^A

A microelectronic system is required to control a mains-operated electrical heater in a commercial greenhouse. The heater:
- should never be on during the day
- should switch on at night when the air temperature drops below a certain level.

Two sensors are required for this system, one to sense whether it is day or night and one to sense the temperature. The potential divider circuits shown in the diagram, **Fig. 7.7**, can be used as inputs. The logic gate should switch on power to the heater when it is cold and dark, so an AND gate is the processor to use. As the heater works from the mains supply, the output device is a relay to switch the heater on and off.

The diagram, **Fig. 7.7**, shows an electronic system that satisfies the requirements.

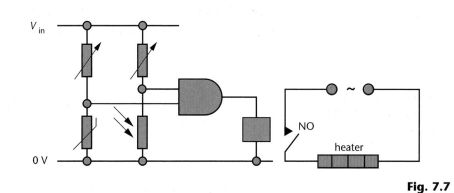

**Fig. 7.7**

## The latch

An electronic system such as a burglar alarm has several sensors that are activated by opening a door or window or movement within a building. Once the alarm has been triggered, it should keep ringing even if the door or window is closed or the burglar stands still. A **latch** is used to maintain the output even if the input from a sensor changes from 1 to 0.

A bistable latch:

- has two stable output states, 0 and 1
- maintains this output when the input signal is removed
- can be set and reset by applying a 1 momentarily to the appropriate input.

The diagram, **Fig. 7.8**, shows a latch which uses two NOR gates.

> The latch is called bistable because it has two output states that are both stable.

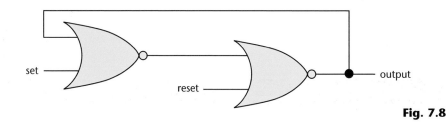

**Fig. 7.8**

A bistable latch can also be made from NAND gates. In the circuit shown in the next diagram, **Fig. 7.9**, the two NAND gates on the left are acting as invertors; if these are omitted the latch is set and reset by applying a momentary 0 rather than a 1.

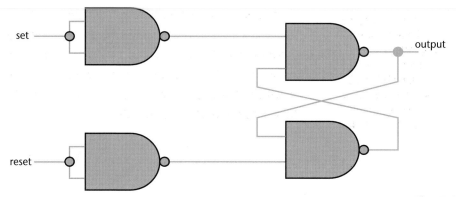

**Fig. 7.9**

## The timing circuit

A **capacitor** is a device that stores charge. It consists of two metal plates separated by an insulator. When connected to a battery or power supply, charge flows onto the plates until the voltage across the capacitor is equal to that across the battery. If a resistor is connected in series the process is slowed down. The diagram, **Fig. 7.10**, shows the charge flow onto a capacitor while it is charging and the way in which the voltage across the capacitor changes with time.

The variation of voltage with time is an exponential rise. When a capacitor discharges, the voltage decreases exponentially. The graph is the same shape as a radioactive decay curve.

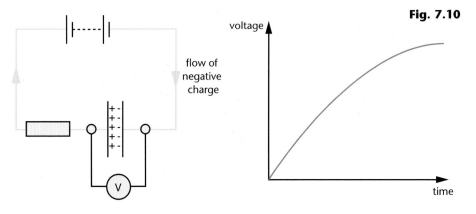

Fig. 7.10

The time taken for the voltage to reach a given value depends on:

Capacitance is measured in farads (F). This is a large unit; the values of practical capacitors are usually measured in microfarads ($1\ \mu F = 1 \times 10^{-6}$ F) or picofarads ($1\ pF = 1 \times 10^{-12}$ F).

- the resistance of the resistor; the greater the resistance, the longer it takes for the capacitor to charge
- the capacitance of the capacitor; the greater the capacitance, the longer it takes for the capacitor to charge.

These factors also have the same effect when a capacitor discharges.

The diagram, **Fig. 7.11**, shows how a potential divider circuit can be made from a capacitor and a resistor. When used as the input to a switching circuit, there is a time delay before the voltage across the capacitor reaches the threshold value required to trigger the circuit into action.

The time delay is determined by the values of the resistor and the capacitor.

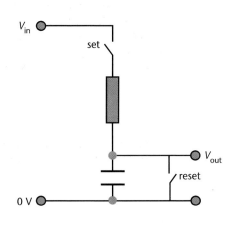

Fig. 7.11

Each time the circuit is used, the capacitor needs to be discharged. This is done by closing the "reset" switch.

## Amplifying the current

If a logic gate circuit is used to drive a relay which requires a higher current than the logic gate can deliver, a **transistor** is used. A small output current from the logic gate is used to switch a much larger current in the transistor, which in turn operates the relay. The diagram, **Fig. 7.12**, shows a light-dependent switch that uses a transistor to switch the current in a relay coil.

> By changing the input sensor, this circuit can be used to switch on devices when there are changes in light level, temperature or humidity.

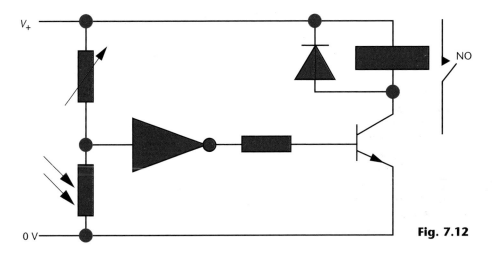

**Fig. 7.12**

In this circuit:

- as the illumination of the LDR is increased, its resistance decreases and the voltage across it drops (voltage across variable resistor increases).
- the **input** to the NOT gate decreases to **below** threshold (a 0) – so the **output** from the NOT gate changes to 1. This starts the transistor conducting, thus operating the relay.
- the variable resistor can be adjusted to change the light level at which the logic gate is switched on
- the diode prevents a large induced voltage when the relay switches off – this would damage the transistor.

---

**PROGRESS CHECK**

1. In a radio, what is:
   a  the input sensor?
   b  the output device?
2. Which logic gate should be used as the processor in a burglar alarm which has two sensors?
3. Explain why a relay is needed for a logic gate to switch a mains-powered appliance.

1a the aerial;  b the loudspeaker;  2 OR;  3 The logic gate operates from a low voltage and can only supply a small current. The mains lamp needs a much higher current at a higher voltage.

# 7.2 Combining resistors

> **LEARNING SUMMARY**
>
> *After studying this section you should be able to:*
>
> ● *calculate the resistance of a number of resistors placed in series and in parallel*
> ● *apply V = I × R to a complete circuit.*

## Resistors in series and in parallel

**NICCEA**
**WJEC**

You should recall from section 1.1 that:

● in a series circuit the sum of the voltages across the individual components is equal to the supply voltage
● components connected in parallel have the same voltage across them
● the resistance of a number of resistors in series is equal to the sum of their individual resistances.

> **KEY POINT**
>
> The last point can be written as
> $$R = R_1 + R_2 + R_3$$
> where $R$ is the total resistance of resistors $R_1$, $R_2$ and $R_3$ connected in series.

The diagram, **Fig. 7.13**, shows resistors connected in series and in parallel.

> A series circuit has only one current path; a parallel circuit has more than one current path.

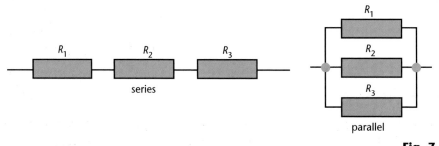

series

parallel

**Fig. 7.13**

Whereas adding resistors in series always increases the resistance, adding them in parallel has the opposite effect. This is because additional resistors in parallel open up more current paths. This results in an increase in current without affecting the voltage.

> The effective resistance of a number of resistors in parallel is always less than the smallest value resistor present.

> **KEY POINT**
>
> The effective resistance of a number of resistors connected in parallel is calculated using the relationship:
> $$\frac{1}{R} = \frac{1}{R_1} + \frac{1}{R_2} + \frac{1}{R_3}$$

When using this relationship:

> A common error is to omit the final step.

● calculate the value of $1/R_n$ for each resistor
● add these together
● calculate the reciprocal of this sum.

## Series and parallel combinations

The diagram, **Fig. 7.14**, shows a combination of both series and parallel resistors.

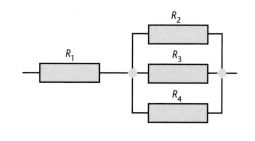

Fig. 7.14

To work out the total resistance, the combination is treated as being effectively two resistors in series:

- first calculate the effective resistance of the three resistors in parallel
- add this to the resistance of $R_1$.

## A complete circuit

The diagram, **Fig. 7.15**, shows a combination similar to that shown above connected to a power supply to form a complete circuit.

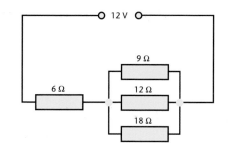

Fig. 7.15

The current in, and voltage across, each resistor can be calculated by applying $V = I \times R$ to the complete circuit and to individual components in the following order:

- calculate the total resistance of the circuit as explained above
- use this to calculate the current that passes in the power supply and the single series resistor
- calculate the voltage across this resistor – the voltage across the parallel combination can now be worked out easily
- when the voltage across the parallel combination is known, the current in each of the parallel resistors can be calculated.

**PROGRESS CHECK**

Calculate the current in, and voltage across, each resistor shown in the circuit above.

| Resistor | Current (A) | Voltage (V) |
|---|---|---|
| 6 Ω | 1.2 | 7.2 |
| 9 Ω | 0.53 | 4.8 |
| 12 Ω | 0.40 | 4.8 |
| 18 Ω | 0.27 | 4.8 |

# Sample GCSE question

1. The diagram shows a logic gate circuit.

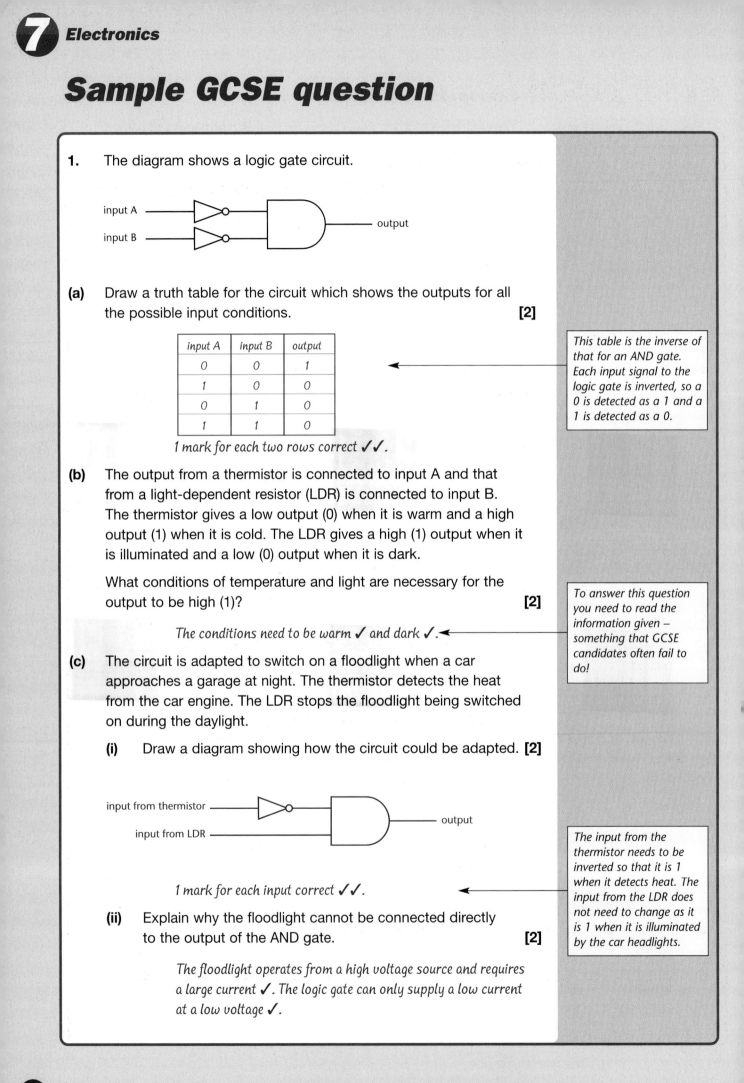

**(a)** Draw a truth table for the circuit which shows the outputs for all the possible input conditions. **[2]**

| input A | input B | output |
|---------|---------|--------|
| 0 | 0 | 1 |
| 1 | 0 | 0 |
| 0 | 1 | 0 |
| 1 | 1 | 0 |

*1 mark for each two rows correct ✓✓.*

*This table is the inverse of that for an AND gate. Each input signal to the logic gate is inverted, so a 0 is detected as a 1 and a 1 is detected as a 0.*

**(b)** The output from a thermistor is connected to input A and that from a light-dependent resistor (LDR) is connected to input B. The thermistor gives a low output (0) when it is warm and a high output (1) when it is cold. The LDR gives a high (1) output when it is illuminated and a low (0) output when it is dark.

What conditions of temperature and light are necessary for the output to be high (1)? **[2]**

*The conditions need to be warm ✓ and dark ✓.*

*To answer this question you need to read the information given – something that GCSE candidates often fail to do!*

**(c)** The circuit is adapted to switch on a floodlight when a car approaches a garage at night. The thermistor detects the heat from the car engine. The LDR stops the floodlight being switched on during the daylight.

**(i)** Draw a diagram showing how the circuit could be adapted. **[2]**

*1 mark for each input correct ✓✓.*

*The input from the thermistor needs to be inverted so that it is 1 when it detects heat. The input from the LDR does not need to change as it is 1 when it is illuminated by the car headlights.*

**(ii)** Explain why the floodlight cannot be connected directly to the output of the AND gate. **[2]**

*The floodlight operates from a high voltage source and requires a large current ✓. The logic gate can only supply a low current at a low voltage ✓.*

# Exam practice questions

1. The diagram shows a logic gate circuit that is part of a car.

   It is designed to operate an alarm if either the driver's or passenger's door is opened while the headlights are switched on.

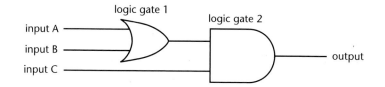

   (a) Write down the names of the logic gates used in this circuit. **[2]**

   (b) Complete the truth table for the circuit. **[4]**

| input A | input B | input C | output |
|---------|---------|---------|--------|
| 0 | 0 | 0 | |
| 1 | 0 | 0 | |
| 0 | 1 | 0 | |
| 1 | 1 | 0 | |
| 0 | 0 | 1 | |
| 1 | 0 | 1 | |
| 0 | 1 | 1 | |
| 1 | 1 | 1 | |

   (c) Which input should be connected to the headlights? **[1]**

   (d) Explain whether the alarm is on when both doors are opened with the headlights switched on. **[2]**

2. The heater in an incubator is controlled by a logic gate. The diagram shows part of the circuit used.

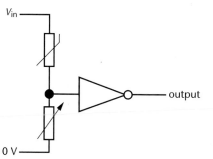

   (a) Explain how the heater is switched on when the temperature of the incubator falls. **[3]**

   (b) What is the advantage of using a variable resistor rather than a fixed resistor? **[1]**

# Exam practice questions

(c)   The input voltage to the circuit is 5.0 V.

At a temperature of 35°C the resistance of the thermistor is 45.0 Ω.

The logic gate switches when the input voltage falls to 2.0 V.

Calculate the resistance of the variable resistor if the heater switches on when the temperature drops to 35°C.   **[3]**

3.   A commercial freezer is fitted with an alarm.

The alarm sounds if the door is left open or the temperature exceeds −10°C.

The diagram shows the system.

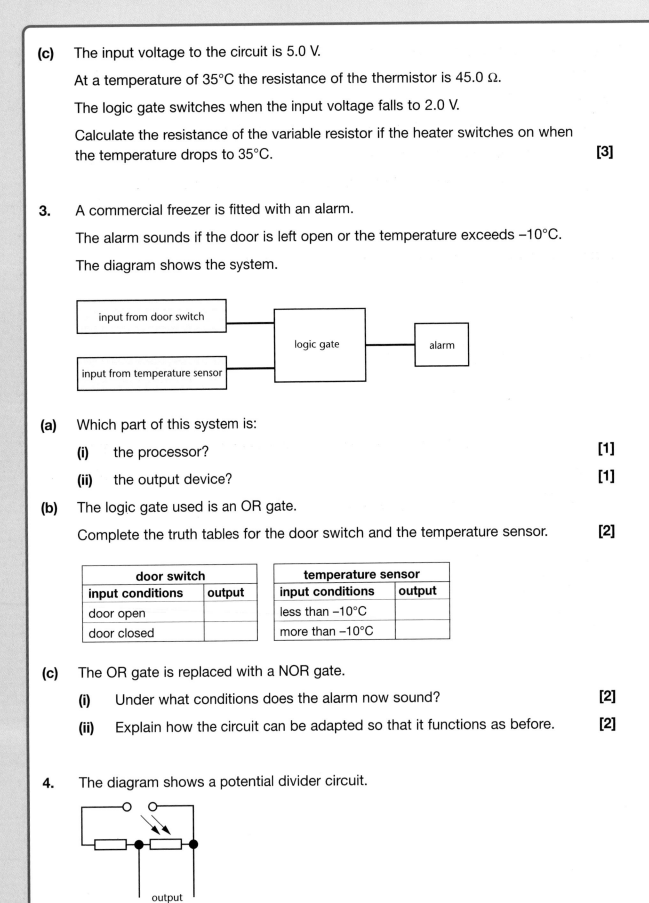

(a)   Which part of this system is:

(i)   the processor?   **[1]**

(ii)   the output device?   **[1]**

(b)   The logic gate used is an OR gate.

Complete the truth tables for the door switch and the temperature sensor.   **[2]**

| door switch | | temperature sensor | |
|---|---|---|---|
| **input conditions** | **output** | **input conditions** | **output** |
| door open | | less than −10°C | |
| door closed | | more than −10°C | |

(c)   The OR gate is replaced with a NOR gate.

(i)   Under what conditions does the alarm now sound?   **[2]**

(ii)   Explain how the circuit can be adapted so that it functions as before.   **[2]**

4.   The diagram shows a potential divider circuit.

# *Exam practice questions*

**(a)** Explain how the output voltage changes when the illumination of the LDR decreases. **[2]**

**(b)** The input voltage to the circuit is 5.0 V.

The output from the LDR is 3.0 V when its resistance is 2000 Ω.

Calculate the resistance of the fixed resistor. **[3]**

**(c)** Explain how the output voltage is affected by increasing the resistance of the fixed resistor. **[2]**

**5.** The diagram shows a circuit that consists of four resistors and a power supply.

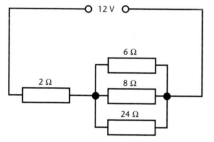

**(a)** Calculate the effective resistance of the three resistors in parallel. **[1]**

**(b)** Complete the table. **[4]**

| resistor | current (A) | voltage (V) |
|----------|-------------|-------------|
| 2 Ω      |             |             |
| 6 Ω      |             |             |
| 8 Ω      |             |             |
| 24 Ω     |             |             |

# 8 Using waves

The following topics are covered in this section:

- **Refraction and lenses**
- **Wave interference**
- **Colour**
- **Resonance**
- **Communicating with waves**

# 8.1 Refraction and lenses

> **LEARNING SUMMARY**
>
> After studying this section you should be able to:
>
> - describe the action of a lens
> - draw a diagram to determine the properties of an image formed by a lens
> - explain how a lens is used in devices such as a magnifying glass and a camera.

## Refractive index

OCR A ᴬ

NICCEA

When light travels from one transparent material into another there is a change of speed. Unless the light is incident along the normal line, this causes a change in direction. The greater the change in speed, the greater the change in direction at a given angle of incidence.

The **refractive index** of a material is a measure of the change in speed that occurs when light passes from a vacuum (or air) into that material.

> **KEY POINT**
>
> refractive index = $\dfrac{\text{speed of light in vacuum}}{\text{speed of light in material}}$

> There is very little difference between the speed of light in a vacuum and in air, so the refractive index can be taken as the ratio of the speed of light in air to the speed in the material.

The greater the refractive index, the greater the change in speed that occurs. The refractive index of water is around 1.3 and that of glass 1.5. When light passes from air into glass there is a greater change of speed than when it passes from air into water.

In a vacuum, light of all wavelengths travels at the same speed. This is not the case when light travels in a material such as water or glass where:

- the shorter the wavelength, the slower the speed
- short-wavelength waves have a greater refractive index and a greater change in speed than longer-wavelength waves as they pass from air into a transparent material
- the change in direction depends on the change in speed, resulting in **dispersion** of white light into different colours.

> Dispersion as light passes through water droplets is the cause of a rainbow.

The dispersion of white light as it passes into glass is shown in the diagram, **Fig. 8.1**.

> This diagram shows that red light is deviated the least when it passes from air into another transparent material, with blue light being deviated the most.

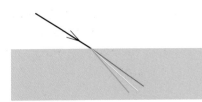

Fig. 8.1

## The action of a lens

When light is emitted from or reflected by an object, it spreads out or **diverges**. The effect of a **concave**, or **diverging**, lens is to increase this divergence while a **convex**, or **converging**, lens decreases it. The diagram, **Fig. 8.2**, shows the shape of these lenses and their action on light.

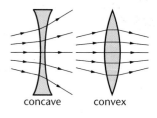

concave    convex

**Fig. 8.2**

The **principal focus** or **focal point** of a lens is the point on the axis through the centre of the lens where:

● the light converges in the case of a convex lens

● the light appears to diverge from in the case of a concave lens.

> **KEY POINT** The focal length, $f$, of a lens is the distance from the centre of the lens to the principal focus.

This is shown in the diagram below, **Fig. 8.3**.

> A convex lens does not always cause light to converge, in some cases it just makes it diverge less.

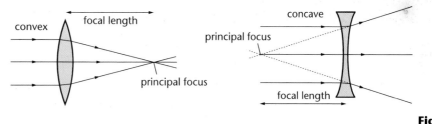

convex    focal length    concave    principal focus

principal focus    focal length

**Fig. 8.3**

## Finding the image

**AQA**
**OCR A** A

Lenses are used to produce images in devices such as telescopes, microscopes and cameras. An image can be **real** or **virtual**. A real image is one that light actually passes through and can be displayed on a screen, for example the image formed on the film in a camera. No light passes through a virtual image and it cannot be displayed on a screen – it only appears to be there. The image that you see when looking in a mirror is a virtual image.

> Images caused by the refraction of light as it passes through a water-air boundary are virtual – an example is the bottom of a swimming pool; it always looks to be nearer than it really is.

Whether an image is real or virtual, and its position and size, can be found by drawing a **ray diagram**. This relies on two key facts:

● light passes through the centre of a lens undeviated

● light incident parallel to the axis passes through the principal focus in the case of a convex lens and appears to have come from the principal focus in the case of a concave lens.

The procedure for drawing a ray diagram is:

● draw a vertical line to represent the lens and a horizontal line to represent the axis through the lens

● to a suitable horizontal scale, draw an upright arrow on the axis to represent the object

> It is easier to draw ray diagrams on squared paper such as graph paper rather than lined or plain paper. The grid enables you to set up a scale and draw parallel lines.

- using the same scale, mark the position of the principal focus on the axis. This should be placed on the same side of the lens as the object for a concave lens, and on the opposite side for a convex lens
- draw a line from the top of the arrow, parallel to the axis until it meets the lens. Continue this line through the principal focus (for a convex lens) or as if it had come from the principal focus (for a concave lens)
- draw a second line from the top of the arrow, straight through the centre of the lens.

If these lines cross after passing through the lens, the image formed is **real**. The top of the image is where the lines cross, and the bottom of the image is the point directly below or above it on the axis. If the lines do not cross, the image is **virtual**. In this case the image position is located by tracing the lines back to find the point where they would both appear to have come from to an observer looking at the light that has passed through the lens.

The position and size of the image can be judged from the horizontal scale. The diagram, **Fig. 8.4**, shows completed ray diagrams for a convex and a concave lens.

> **Dotted lines are used here for the path that light appears to have followed and to represent a virtual image.**

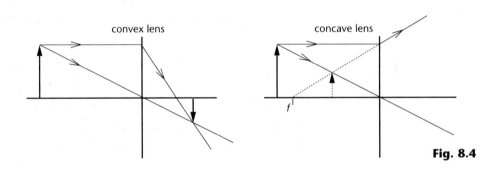

convex lens    concave lens

**Fig. 8.4**

## Types of image

> **If the object is placed precisely at $f$ no image is formed – the light that emerges from the lens is parallel.**
>
> **If the object is placed precisely at $2f$, the image is real, inverted and the same size as the object.**

A concave lens on its own always forms an image that is virtual, upright and smaller than the object. Convex lenses can form three types of image, depending on the distance between the object and the lens:

- if the object is between the lens and a distance equal to its focal length, $f$, the image is virtual, upright and magnified – this corresponds to using the lens as a **magnifying glass**
- if the object is between $f$ and $2f$ from the lens the image is real, upside down and magnified – this corresponds to using the lens as a **projector**
- if the object is further away than $2f$, the image is real, upside down and smaller than the object – this corresponds to using the lens as a **camera**.

## Optical instruments

To use a convex lens as a **magnifying glass**, you should look through the lens at the object. The diagram, **Fig. 8.5**, shows that the light that has passed through the lens is less divergent than when it entered the lens. This produces a magnified image that is further away than the object. The closer the object is to the principal focus, the greater the magnification.

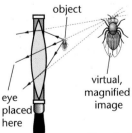

object

virtual, magnified image

eye placed here

**Fig. 8.5**

A **projector** and a **camera** both produce real images; that from the projector is magnified but the image that forms on the film in a camera is usually smaller than the object being photographed. The diagrams, **Fig. 8.6**, show the action of a projector and a camera.

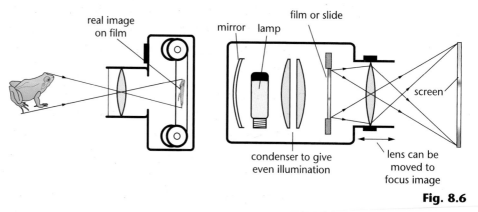

**Fig. 8.6**

To focus these instruments to make the image on the screen sharp:

● the camera lens is moved away from the film to focus on a near object, and towards the film to focus on a close object

● the projector lens is moved towards the slide to increase the magnification when the screen is moved away from the projector.

## Sight

The eye uses two convex lenses, one fixed and one variable, to produce a real image. The diagram, **Fig. 8.7**, shows some key components of the eye.

> The fixed lens is the cornea, which does most of the focusing. This is then "fine-tuned" by the eye lens.

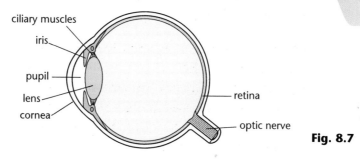

**Fig. 8.7**

> The cells that are sensitive to brightness are called rods and those sensitive to colour are called cones. Cones do not function in low light levels, so you cannot tell the colour of grass by moonlight!

● the **retina** contains cells that are sensitive to the brightness and colour of light

● electrical impulses from these cells are sent along the **optic nerve** to the brain

● the **iris** is a circular piece of tissue that controls the size of the **pupil**

● the **ciliary muscles** adjust the shape of the **lens** to focus on near and distant objects; this process is called **accommodation**.

Two common reasons for wearing spectacles or contact lenses are:

● **short sight**, where near objects can be seen clearly but the light from a distant object forms a real image in front of the retina

● **long sight**, where distant objects can be seen clearly but light from a near object is not brought to a focus in the eyeball.

The diagram, **Fig. 8.8**, shows the effects of short and long sight and how they can be corrected using lenses.

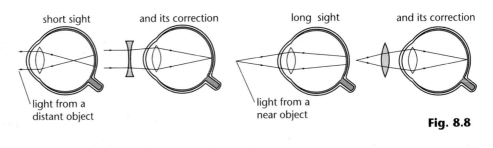

short sight     and its correction     long sight     and its correction

light from a distant object

light from a near object

**Fig. 8.8**

**PROGRESS CHECK**

1. The speed of light in air is $3.00 \times 10^8$ m/s.
   The refractive index of glass is 1.50.
   Calculate the speed of light in glass.
2. Whereabouts should an object be placed relative to the lens when a convex lens is used as a magnifying glass?
3. A 2 cm × 2 cm slide is used in a slide projector.
   The slide is placed 12.0 cm from the projection lens, which has a focal length of 10.0 cm. Use a ray diagram to find how far the screen should be placed from the lens?

1. $2.00 \times 10^8$ m/s;    2. Between the lens and its principal focus;    3. 60 cm.

# 8.2 Resonance

**LEARNING SUMMARY**

*After studying this section you should be able to:*

- *recall the meaning of natural frequency*
- *explain how resonance occurs*
- *describe the modes of vibration of a string.*

## Natural vibrations

OCR A ᴬ    OCR A ᴮ

Every object has its own **natural frequency** of vibration, known as its **characteristic frequency**. This is the frequency of the oscillation when it is displaced and allowed to vibrate freely. Plucking a rubber band or pushing a swing causes a vibration at the natural frequency.

The diagram, **Fig. 8.9**, shows two systems whose natural frequency of vibration can be changed.

> The natural frequency of a rubber band can be changed by altering the tension.

> The stiffness of a spring is a measure of how difficult it is to stretch it.

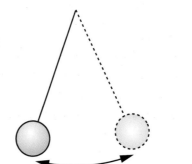

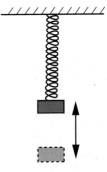

The natural frequency of a pendulum depends only on the length of the string, but that of a mass on a string depends on the mass and the stiffness of the spring

**Fig. 8.9**

All objects can also be forced to vibrate by being in contact with another vibrating object, but in many cases the amplitude of a **forced vibration** is small.

> **KEY POINT**
> If the frequency of a forced vibration is the same as the natural frequency, then a large amplitude vibration builds up. This is called resonance.

Examples of resonance include:

- pushing a child on a swing – the amplitude increases when the swing is pushed at a frequency equal to its natural frequency
- rattling exhausts – parts of a motor vehicle can vibrate with a large amplitude when their natural frequency matches that of the engine vibrations
- playing a clarinet – a column of air is forced to vibrate by a vibrating reed.

*Electrical resonance is used to tune radios and televisions to different stations.*

Blowing across a column of air, such as in an empty or partly-filled milk bottle, can also cause it to vibrate at its natural frequency. This is how a church organ produces notes. Different notes are played by pipes of different sizes – long pipes have lower natural frequencies than shorter ones.

## Vibrating strings

Pianos, guitars and violins all use the vibrations of a string in tension to produce sound. Strings can vibrate in a number of ways, called **modes**. The diagram, **Fig. 8.10**, shows the vibrations of a string at its natural frequency. This is known as the **fundamental mode** of vibration.

*In the fundamental mode the length of the vibrating string is half the wavelength of the sound wave produced.*

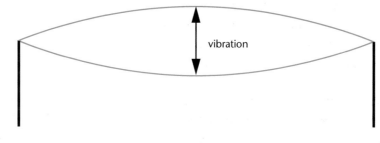

vibration

**Fig. 8.10**

In this mode of vibration:

- the ends of the string are **nodes** – points of no displacement
- there is an **antinode** – where the maximum displacement occurs – in the centre of the string.

The frequency of the fundamental mode of vibration of a string depends on:

- its length – increasing the length reduces the natural frequency
- the mass per unit length – the greater this is, the lower the natural frequency
- the tension – increasing the tension increases the natural frequency.

*The lowest frequency notes on a piano are produced by the longest strings – these also have a greater mass per unit length than the strings used to play the high notes.*

Strings can also vibrate with large amplitudes at frequencies equal to a whole number multiple of the natural frequency. The diagram, **Fig. 8.11**, shows the modes of vibration of a string at frequencies equal to twice and three times the natural frequency.

> The fundamental mode of vibration has just one antinode, but other modes have several.

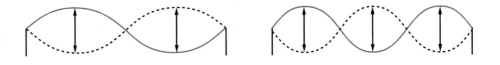

**Fig. 8.11**

When a stringed instrument is played, the strings vibrate in several modes at once. This is why the same note played on different instruments has a different sound. The **quality** of a note depends on the number and relative intensity of the different modes of vibration. The displacement–time graphs, **Fig. 8.12**, compare the "pure" note from a tuning fork with the same note played on a piano.

> Analysis of the note from a tuning fork shows that it is a sine wave. A signal generator can also be used to produce a "pure" note.

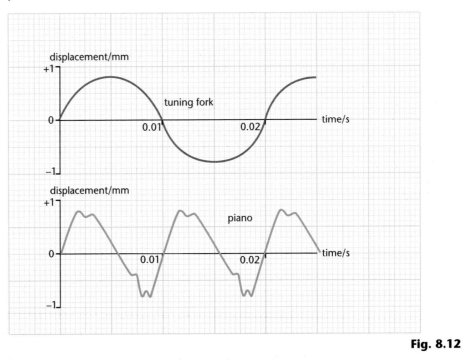

**Fig. 8.12**

A displacement–time graph can be used to determine:

- the amplitude of the wave – the maximum displacement
- the frequency of the wave by measuring the time, $T$, for one complete cycle and using the relationship $f = 1/T$
- the quality of the note by judging how much it deviates from a "pure" note.

---

**PROGRESS CHECK**

1. Suggest why a washing machine can vibrate with a large amplitude as the speed of the spin increases.
2. What is required for an object to resonate?
3. What affects the quality of the note played on a violin?

1. The vibrations of the drum are at the natural frequency of the washing machine;  2. A forced vibration at the same frequency as the natural frequency;  3. The number and intensity of the modes of vibration present.

# 8.3 Wave interference

**LEARNING SUMMARY**

After studying this section you should be able to:

● recall that interference involves cancellation and reinforcement of waves
● describe interference effects with sound, surface water waves and electromagnetic waves
● explain how constructive and destructive interference are related to path difference.

## What is interference?

OCR A ᴬ
WJEC

**Interference** is a property of all wave motions. It results in the **cancellation** or **reinforcement** of waves when the paths of two waves of the same type cross. The effects of interference are readily observed using sound but are more difficult to observe with short-wavelength waves such as light. The diagram, **Fig. 8.13**, shows how an observer can detect the effect of sound waves interfering.

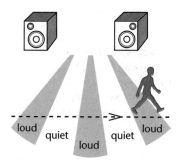
superposition of sound waves
**Fig. 8.13**

The observer notices that:

● there are places where the sound is louder than that due to one loudspeaker alone
● there are places where the sound is almost inaudible.

> The separation of the loud and quiet sounds depends on three factors; the wavelength of the sound, the separation of the loudspeakers and the distance between the sources and the observer.

This is due to **constructive interference** and **destructive interference** taking place between the waves from the two sources.

> **KEY POINT**
>
> Constructive interference results in a loud sound. It occurs at any point where the displacements from both waves are in the same direction. Destructive interference occurs at points where the displacements are in opposite directions and results in a quiet sound.

The diagram, **Fig. 8.14**, shows how two waves interfere constructively and destructively when they cross.

> The waves only show complete cancellation if they have the same amplitude.

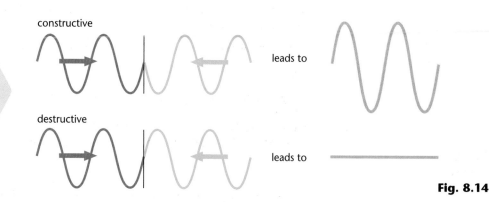

constructive

leads to

destructive

leads to

**Fig. 8.14**

## Interference on water

Interference effects can also be noticed with surface water waves. The diagram, **Fig. 8.15**, shows what happens when two dippers create circular waves on water. The solid lines represent wave crests and the broken lines represent wave troughs.

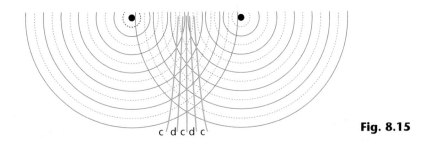

c  d  c  d  c

**Fig. 8.15**

> The factors that affect the separation of the lines of constructive and destructive interference are the same as those for the sound waves.

Where two wave crests or two wave troughs cross they interfere constructively – some lines of constructive interference (c) have been marked on the diagram. Destructive interference takes place at points where a wave crest meets a trough. Lines marked (d) show destructive interference.

## Path difference

All electromagnetic waves also show interference effects. The diagram, **Fig. 8.16**, shows an arrangement for demonstrating interference of 3 cm microwaves.

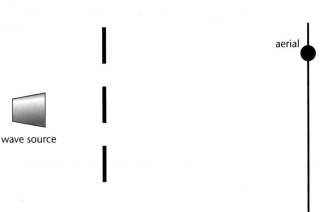

aerial

wave source

**Fig. 8.16**

> The gaps in the barrier should have a width comparable to the wavelength for adequate spreading to take place.

In this case two copies of a single wave source are obtained by **diffraction** through the two gaps in the barrier. These then interfere and variations in intensity are detected by an aerial. At a point equidistant from both gaps the waves interfere constructively. Since the waves have travelled along equal-length paths, the same signal arrives from each wave source.

At all other points, there is a difference in the length of the paths travelled by the waves; this is known as the **path difference**. If this is equal to a whole number of wavelengths the same signal arrives from each wave and the interference is constructive. If the path difference is equal to half a wavelength or one and a half wavelengths then the signals from the two waves have opposite displacements, so they interfere destructively.

This is shown in the diagram, **Fig. 8.17**.

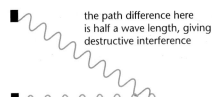

the path difference here
is half a wave length, giving
destructive interference

the path difference here
is one wave length, giving
constructive interference

**Fig. 8.17**

This is true provided
that the two
wave sources are
vibrating in step
with each other.

**KEY POINT**

**Constructive interference** occurs when the path difference between
waves from two sources is equal to a whole number of wavelengths,
or an even number of half wavelengths.
**Destructive interference** occurs when the path difference is equal to
an odd number of half wavelengths.

Interference is more difficult to observe with light due to its shorter wavelength.
The fact that it does happen provides evidence for light having a wave-like
nature.

**PROGRESS CHECK**

1. Explain what is meant by *destructive interference*.
2. How is *diffraction* used in the demonstration of interference between two sources
   of 3 cm electromagnetic waves?
3. What is the condition for waves from two sources vibrating in step to interfere
   constructively?

1. The cancellation of two waves with opposite displacements;   2. To produce two identical
copies of a single wave source;   3. The path difference should be a whole number of wavelengths, or an even number of half wavelengths.

# 8.4 Communicating with waves

**LEARNING SUMMARY**

*After studying this section you should be able to:*

● describe how information is transmitted as analogue and digital signals
● explain how the transmission of radio waves through the atmosphere depends on the wavelength of the waves
● describe some uses of satellites.

## Systems

Edexcel A  Edexcel B

Communications using sound, light and radio may appear to be very different processes, but there are a number of features common to many communication systems:

● a **transducer** transfers a signal from one form to another
● an **encoder** translates the information into a form that can be transmitted
● a **modulator** produces periodic variations in amplitude and/or frequency of the wave that carries the coded signal
● an **amplifier** increases the amplitude of a signal
● a **transmitter** sends the signal from one place to another
● a **receiver** detects a transmitted signal
● a **demodulator** separates a coded signal from the wave that carries it
● a **decoder** translates the information from the form in which it is transmitted to one in which it can be understood.

> Sound is modulated by the mouth during speech.

### Transmitting along wires and fibres

Signals that carry telephone conversations, and radio and television programmes spend some part of their journey from the transmitter to the receiver travelling along wires or fibres or both. In each case:

> All signals are subject to attenuation as they travel.

● the signals become **attenuated** due to loss of energy – this results in them becoming weaker
● the signals pick up **noise** which affects the amplitude of the wave and results in distortion of the signal.

> Electrical signals travelling in wires need to be repeated every few miles. Optical signals travelling along fibres have a much greater range.

Most optical fibre transmissions use digital signals, where the information is encoded as a series of pulses of light or infra-red radiation. Because of its higher frequency, light can carry more information that radio waves. Both analogue and digital signals need to be amplified at several stages during their transmission because of the effects of attenuation. Digital signals can be restored to their original condition in a process known as **regeneration** (see section 3.2). Analogue signals are amplified by repeaters, which amplify both the signal and the noise. This is shown in the diagram, **Fig. 8.18**.

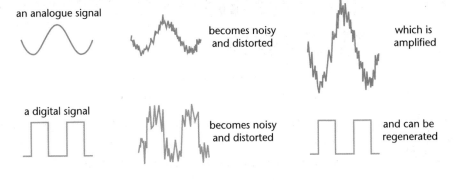

an analogue signal

becomes noisy and distorted

which is amplified

a digital signal

becomes noisy and distorted

and can be regenerated

**Fig. 8.18**

# Storing and reading information

Edexcel A    Edexcel B
OCR A B

Computer disks store information in a **digital** form – each **bit** of information can be a 0 or a 1. This information is stored in a magnetic form. A **compact disc** (CD) also stores information digitally as a series of "pits" and "bumps" on its surface. The diagram, **Fig. 8.19**, shows how a **diode laser** and a **photodiode** are used to read the information on a compact disc.

**The photodiode detects light from the diode laser that has been reflected at the surface of the CD.**

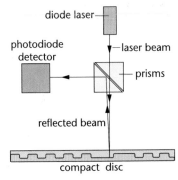

**Fig. 8.19**

A **vinyl record** stores information in an **analogue** form – here the information is stored in grooves that have both horizontal and vertical bumps. The bumps vary in frequency and amplitude in the same way as the sound recorded.

**A domain is a magnetised region on the tape.**

**Magnetic tape** can be used to store either type of signal by rearranging the magnetic **domains** on the recording tape. On an unrecorded tape, these are arranged randomly, but when a signal is recorded these domains are ordered in a way that represents the information either as a varying magnetic field for an analogue signal or a series of magnetic pulses in the case of a digital signal.

Information is written onto and read from magnetic tape using a tape head similar to that shown in the diagram, **Fig. 8.20**.

There are three types of tape head in a tape recorder:

**Fig. 8.20**

**In some inexpensive tape recorders the same tape head is used for recording and playback.**

- the record head writes information onto the tape when an alternating current passes in the coil and produces an alternating magnetic field in the gap that the tape passes over
- the erase head re-scrambles the magnetic domains when a high frequency alternating current passes in the coil; this removes any recorded information
- the playback head reads information by electromagnetic induction – an alternating voltage is induced in the coil due to the changing magnetic field of the moving tape.

The record and playback heads are examples of **transducers**; they transfer information from one form to another.

## Other key transducers

A **microphone** transfers the information from a signal in the form of sound to an electrical signal. One common form of microphone is the **moving coil microphone**, shown in the diagram, **Fig. 8.21**. In this transducer:

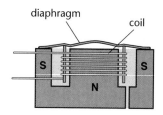

diaphragm

coil

**Fig. 8.21**

- the sound wave causes vibration of the diaphragm
- this causes the coil to vibrate in the radial magnetic field of a cylindrical magnet
- a voltage is induced in the coil with a variation in frequency and amplitude similar to that of the sound.

In a **loudspeaker** the process is reversed – information from an electrical signal is reproduced as a sound. The diagram, **Fig. 8.22**, shows a **moving coil loudspeaker**. In this transducer:

cylindrical magnet

paper cone

**Fig. 8.22**

- an alternating current that represents the signal passes in the coil
- interaction of the magnetic fields due to the current and the fixed magnet produces a force on the coil
- the coil vibrates, pushing the paper cone in and out.

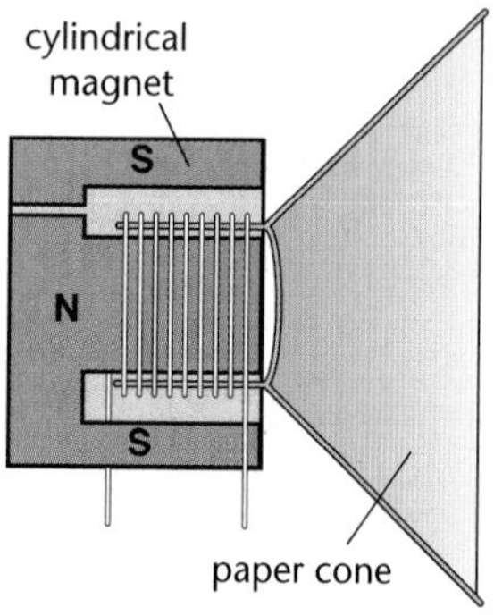

# Using radio waves

Prior to the development of radio, long-range communication depended on electric currents and needed a wire link from the transmitter to the receiver. In the mid-nineteenth century the **telegraph** system was used to send messages using Morse code and Alexander Graham Bell's invention of the telephone in 1876 enabled people to be able to talk directly to each other. At the beginning of the twentieth century radio was beginning to be developed and was the first "wireless" means of long-distance communication. More recent developments include the use of satellites which have made world-wide communications possible.

Radio waves can travel in a number of different ways, depending on their wavelength and frequency:

- long-wavelength waves with frequencies up to 2 MHz travel as **ground waves** – they follow the curvature of the Earth
- medium-wavelength waves with frequencies in the range 3–30 MHz are reflected by the ionosphere, a layer in the upper atmosphere, and the ground – they are called **sky waves**
- short-wavelength waves with frequencies above 30 MHz travel in straight lines and can pass through the ionosphere – these are called **space waves** as they can travel through the atmosphere and out to space.

# Using satellites

Some early satellites used for a radio transmission merely acted as reflectors of radio waves – they were **passive satellites**. An **active satellite** receives, amplifies and re-transmits signals. Artificial satellites in orbit around the Earth are used for a number of purposes:

> **Satellite navigation systems can now be used to track the movement of a stolen vehicle.**

- **navigation** – a ship or a car can use satellite transmissions to determine its position with a precision of a few metres
- **exploration** of the Solar System and the Universe – satellites such as the Hubble telescope can detect radiation that is absorbed by the Earth's atmosphere
- **surveillance** – the movement of armies and weapons can be monitored by satellites
- **weather** – the movement of clouds and weather fronts is monitored by satellites
- **communications** – telephone conversations and television broadcasts are transmitted around the world by satellites.

The time it takes for a satellite to complete one orbit of the Earth depends on the height of the satellite. Satellites occupy different orbits according to the job that they do.

Some weather satellites occupy **low polar orbits** similar to that shown in the diagram, **Fig. 8.27**. In a low polar orbit:

> **The Earth rotates through an angle of 24° during one orbit of a satellite in a low polar orbit.**

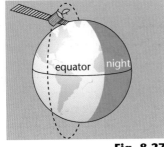

- a satellite completes an orbit of the Earth in 96 minutes
- during this time the Earth spins on its axis, so the satellite "sees" a different part of the Earth on the next orbit
- the whole of the Earth's surface is monitored during one day.

**Fig. 8.27**

For telecommunications, satellites that do not move relative to the Earth are often needed. These occupy **geostationary orbits**. A satellite in a geostationary orbit:

- completes one orbit of the Earth in 24 hours
- remains in the same position above the Earth's surface
- orbits above the equator.

---

**PROGRESS CHECK**

1. Which part of a domestic radio is the receiver?
2. How is information encoded before being transmitted along an optical fibre?
3. Explain why satellite transmissions use very short wavelength radio waves.

1. The aerial;  2. As pulses of light or infra-red radiation;  3. Long wavelength waves cannot pass through the ionosphere.

# 8.5 Colour

**LEARNING SUMMARY**

**After studying this section you should be able to:**

- describe the difference between a primary colour and a secondary colour
- explain the effect of colour filters
- predict the appearance of coloured objects in different colours of light.

## Primary and secondary colours

WJEC

The visible part of the electromagnetic spectrum consists of an uncountable number of colours ranging from blue to red. In a colour television and computer monitor all of these colours (and some additional ones) are produced from a combination of just three colours: red, green and blue.

**KEY POINT**
Red, green and blue are called the primary colours because any other colour of light can be obtained by combinations of these three colours.

> Mixing coloured lights is colour addition. It is not like mixing paints which, like colour filters, subtract colours.

The colour obtained by combining two primary-coloured lights of equal intensity is called a **secondary colour**. The diagram, **Fig. 8.28**, shows the results of overlapping beams of light of the three primary colours on a white screen. This diagram shows that:

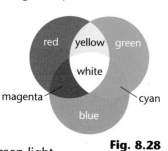

- yellow light is produced by mixing red light and green light
- magenta light is produced by mixing red light and blue light

> Secondary colours are used in colour printing and colour photographs.

- cyan light is produced by mixing blue light and green light

**Fig. 8.28**

- white light is produced by mixing all three primary colours of light.

### Colour filters

Colour filters remove some colours from white light, and allow others to pass through. A primary colour filter allows light of its own colour only to pass through, but that of a secondary colour transmits its own colour as well as the two primary colours that combine to make the secondary colour. The table shows the primary colours of light transmitted and absorbed by different coloured filters.

> Colour filters are used in theatres to produce dramatic effects.

| Colour of filter | Primary colours transmitted | Primary colours absorbed |
|---|---|---|
| red | red | green and blue |
| green | green | red and blue |
| blue | blue | red and green |
| yellow | red and green | blue |
| magenta | red and blue | green |
| cyan | green and blue | red |

When two filters are combined each one absorbs certain colours, so the light that emerges contains the colours that are not absorbed by either filter. The diagram, **Fig. 8.29**, shows what happens when white light passes through a yellow filter followed by a magenta one.

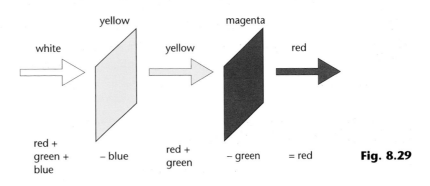

Fig. 8.29

## Coloured objects

Light sources are the exception to this – they are seen by the light that they give out.

Most objects are seen by the light that they reflect. A white surface reflects all colours of light and a black surface absorbs all colours and reflects none. Other colours of surface are more selective – some colours are absorbed and some are reflected. The colours that any particular object absorbs and reflects can be deduced from its appearance in white light, since coloured objects act in a similar way to colour filters.

Objects look to be different colours in coloured lights because some of the colours that they normally reflect are not present. To predict what colour an object looks to be when coloured light is shone on it:

This does not work for objects viewed in yellow street lighting. The light from these lamps is pure yellow, not a mixture of red and green.

- write down the primary colours present in the coloured light
- cross out the primary colours that the object absorbs
- if no primary colours are left, the object appears black. If one primary colour is left it appears to be that colour and if two primary colours are left it appears to be the secondary colour produced when light of those two primary colours is combined.

The diagram, **Fig. 8.30**, shows the effect of illuminating a yellow tee-shirt with magenta coloured light.

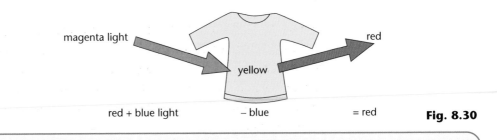

Fig. 8.30

1. Which two primary colours of light combine to form (a) cyan, (b) magenta and (c) yellow?
2. Explain why no light passes through a combination of two primary colour filters.
3. A top appears cyan in white light.
   What colour does it appear in yellow light?

1a. blue and green; b. red and blue; c. red and green;    2. The primary colour transmitted by the first filter is absorbed by the second;    3. Green.

# Sample GCSE question

**1. (a)** A telephone conversation can be sent either as an analogue signal along a wire or as a digital signal along an optical fibre.

Describe the difference between an analogue and a digital signal. **[2]**

*An analogue signal varies continuously ✓.*
*A digital signal can only have certain values, usually 0 or 1 ✓.*

> Always be explicit when answering questions in tests and examinations. It is better to give too much information than too little, since extra information is not penalised.

**(b)** Digital signals can be *regenerated*. Explain what this means. **[3]**

*The signal is amplified ✓ and noise is removed ✓ to return the signal to its original condition ✓.*

**(c)** Suggest two advantages of recording sound as a digital signal rather than an analogue signal. **[2]**

*Noise can be removed from a recording, so there is less distortion ✓.*
*More information can be stored in the same space ✓.*

> The cue word "suggest" means that this is not required knowledge. You are expected to use knowledge of similar situations to answer this question.

**(d)** The diagram shows the earpiece of a telephone.

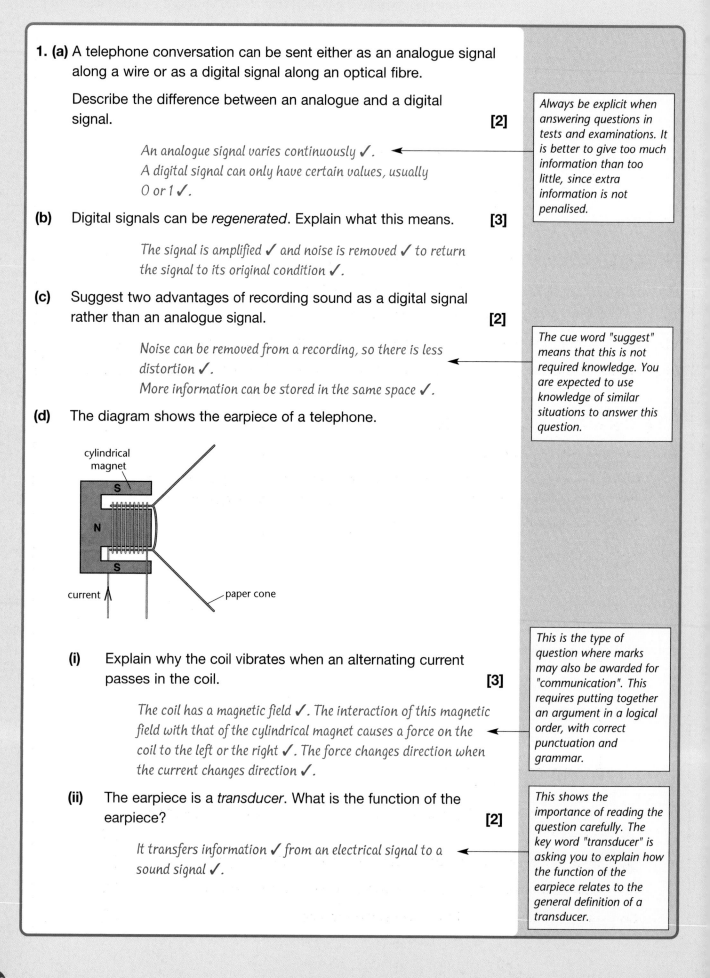

cylindrical magnet

S

N

S

current

paper cone

**(i)** Explain why the coil vibrates when an alternating current passes in the coil. **[3]**

*The coil has a magnetic field ✓. The interaction of this magnetic field with that of the cylindrical magnet causes a force on the coil to the left or the right ✓. The force changes direction when the current changes direction ✓.*

> This is the type of question where marks may also be awarded for "communication". This requires putting together an argument in a logical order, with correct punctuation and grammar.

**(ii)** The earpiece is a *transducer*. What is the function of the earpiece? **[2]**

*It transfers information ✓ from an electrical signal to a sound signal ✓.*

> This shows the importance of reading the question carefully. The key word "transducer" is asking you to explain how the function of the earpiece relates to the general definition of a transducer.

# Exam practice questions

**1.** The diagram represents an object placed at a distance of 8.0 cm from a convex lens of focal length 5.0 cm.

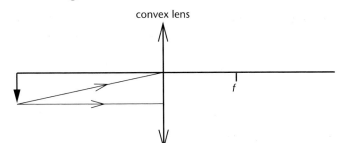

convex lens

f

**(a)** Complete the ray diagram and draw in the image. **[3]**

**(b)** What is the distance between the lens and the image? **[1]**

**(c)** State two differences between the object and the image. **[2]**

**(d)** Which optical instrument uses a convex lens to produce this type of image? **[1]**

**(e)** Explain why the object is normally placed upside-down in this optical instrument. **[1]**

**2. (a)** What is meant by the *natural frequency* of vibration of an object? **[1]**

**(b)** Explain how *resonance* occurs. **[2]**

**(c)** A metal wire is held in tension. The tensile force remains unchanged while the length of the wire is varied. The graph shows how the natural frequency of the wire changes as its length is increased.

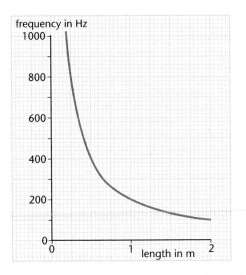

**(i)** What is the relationship between the natural frequency and the length of the wire? **[2]**

**(ii)** What is the natural frequency when the wire is 0.80 m long? **[1]**

**(iii)** What is the wavelength of the sound produced at this frequency? **[1]**

**(iv)** Calculate the speed of the wave along the wire. **[3]**

# Exam practice questions

**3.** The diagram shows how an interference pattern can be observed when two beams of light overlap.

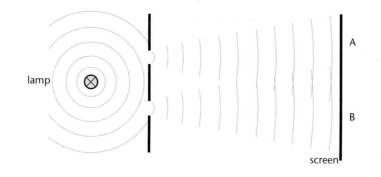

**(a)** Name the effect that occurs when the light passes through the narrow gaps. **[1]**

**(b)** Explain what is seen on the screen at A. **[2]**

**(c)** Explain what is seen on the screen at B. **[3]**

**(d)** What is meant by *path difference*? **[1]**

**(e)** When two wave sources are emitting waves in step, what is the path difference at points where constructive interference takes place? **[1]**

**4.** Magnetic tape can be used to store analogue or digital signals.

**(a)** Which type of signal is stored on a compact disc? **[1]**

**(b)** Which type of signal is supplied to a loudspeaker? **[1]**

**(c)** The diagram shows a tape head being used to record a signal on a tape.

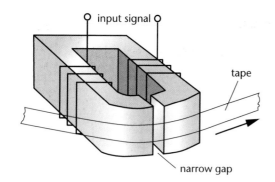

**(i)** In what form is the signal stored on the tape? **[1]**

**(ii)** Describe the changes that take place as the tape passes over the narrow gap. **[2]**

**(iii)** A similar tape head is used to play back the recorded signal.

Explain how it does this. **[3]**

# Exam practice questions

**5.** The diagram shows how a dish aerial is used to focus radio waves before they are transmitted to a satellite.

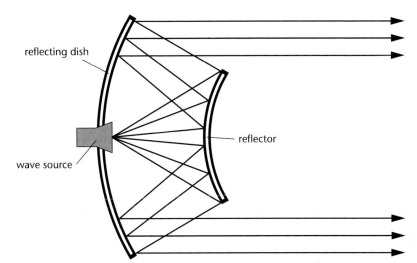

reflecting dish

reflector

wave source

**(a)** Explain how focusing the beam reduces attenuation. **[2]**

**(b)** Radio waves can travel as ground waves, sky waves or space waves.

Explain which type of wave is used for satellite transmissions. **[2]**

**(c)** The satellite is in a geostationary orbit.

    **(i)** What are the features of a geostationary orbit? **[2]**

    **(ii)** What is the advantage of using a geostationary orbit for a telecommunications satellite? **[1]**

**(d)** The waves are received by an active satellite. What does an active satellite do? **[3]**

**(e)** The satellite transmission uses digital signals.

Explain why these are preferred to analogue signals. **[3]**

**(f)** Suggest three reasons why very short wavelength radio waves are used for the satellite transmission. **[3]**

**The following topics are covered in this section:**

- **Projectiles and momentum**
- **Turning in a circle**
- **Some effects of forces**

# 9.1 Projectiles and momentum

**LEARNING SUMMARY**

*After studying this section you should be able to:*

- **use the equations of motion to solve problems**
- **describe the motion of a projectile**
- **explain how the principle of conservation of momentum applies to collisions.**

## Equations of motion

OCR A ^A   OCR A ^B
NICCEA
WJEC

When an object moves at a constant speed, it travels equal distances in equal time intervals. An accelerating object travels increasing distances in successive equal time intervals. If the acceleration is constant the relationships between the variables are described by the **equations of motion**. The variables are:

- the **initial velocity**, $u$
- the **final velocity**, $v$
- the **time** of motion, $t$
- the **displacement**, $s$
- the **acceleration**, $a$.

> A constant acceleration is also known as a uniform acceleration.

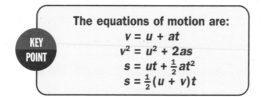

**KEY POINT**

The equations of motion are:
$$v = u + at$$
$$v^2 = u^2 + 2as$$
$$s = ut + \tfrac{1}{2}at^2$$
$$s = \tfrac{1}{2}(u + v)t$$

> If a ball is thrown vertically into the air, when it returns to its starting point the displacement is zero, but it has travelled some distance.

Each equation is a relationship between four variables, so if the values of three are known the other two can be calculated. When using these equations it is important to remember that:

- $s$ represents displacement; this may not have the same value as the distance travelled (see p 25)
- $v$, $u$, $a$ and $s$ all have direction as well as size, so + and – signs should be used if, for example, the acceleration is in the opposite direction to the velocity.

### Projectiles

When a ball is thrown horizontally, it starts to accelerate vertically as soon as it leaves the hand. If the effects of air resistance are negligible then its horizontal speed is constant. Different relationships apply to the horizontal and vertical motions and they should always be analysed independently.

> **KEY POINT**
>
> The **equations of motion** apply to the vertical motion.
> The relationship average speed = distance travelled ÷ time taken applies to the horizontal motion.

The common factor between the horizontal and vertical motion of a projectile is the time for which it is in the air.

Objects which are projected either horizontally or at the same angle to the horizontal are called **projectiles**. They all follow the same pattern of motion:

- horizontally, a projectile travels equal distances in equal time intervals
- vertically, a projectile travels increasing distances in successive equal time intervals.

The path of a projectile is shown in the diagram, **Fig. 9.1**. The steepness of the curve depends on the horizontal speed.

In accelerated motion, the distance travelled is proportional to the square of the time.

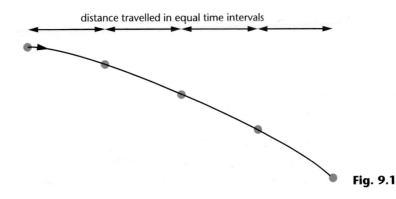

distance travelled in equal time intervals

Fig. 9.1

## Adding physical quantities

OCR A B
NICCEA

What is the result of adding 2 kg to 2 kg? There is only one possible answer to this question, 4 kg. But if the question is changed slightly to "what is the result of adding together two 2 N forces?" the answer could be any value between 0 N and 4 N. This is because, unlike mass, forces have a **direction**. Quantities that have size only are called **scalars**, but those that also have a direction are **vectors**, and different rules of addition apply. The table gives some examples of vector and scalar quantities.

The sum of two vector quantities is sometimes called the resultant.

| vectors | scalars |
|---|---|
| force | mass |
| displacement | distance |
| velocity | speed |
| acceleration | temperature |

If two vector quantities of the same type act in parallel on the same object, working out their effect is a case of simple addition or subtraction. This is shown in the diagram, **Fig. 9.2**.

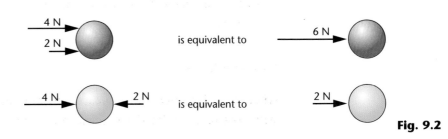

Fig. 9.2

To add together two vector quantities acting at right angles to each other, the following procedure is used:

● to a linear scale, draw an arrow that represents one of the vectors in size and direction

● starting at the end of the first arrow, draw a second arrow that represents the other vector

● a line drawn from the beginning of the first vector to the end of the second vector represents their sum, in size and direction.

The diagram, **Fig. 9.3**, shows the application of this method to find the sum of two forces, 5 N and 3 N, acting at right angles on the same object.

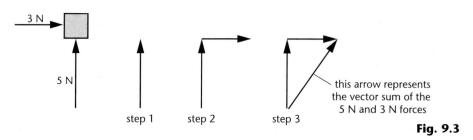

**Fig. 9.3**

In this case the sum of these vectors is a force of 5.8 N acting at an angle of 31° to the 5 N force.

## What is momentum?

Why is a heavy, slow-moving ball used in ten-pin bowling rather than a lighter, faster-moving ball? The heavier ball is more effective because it has more **momentum**.

Momentum is a useful concept for predicting the outcome of a collision. The momentum of a moving object depends on:

● its mass

● its velocity.

**KEY POINT**

**momentum = mass × velocity**
$$p = m \times v$$
**Momentum is a vector quantity and is measured in kg m s$^{-1}$ or the equivalent unit N s.**

When two objects collide there is no overall change in momentum, provided that the only forces acting are those between the objects themselves. The forces between the objects are equal in size and opposite in direction, and so are their changes in momentum. Momentum is always conserved – this is known as the **principle of conservation of momentum**.

## Types of collision

AQA
OCR A ᴬ    OCR A ᴮ
NICCEA
WJEC

When two "vehicles" collide on an air track the external resistive forces are small and so the principle of conservation of momentum can be used to predict the outcome. The fact that velocity and momentum have direction as well as size needs to be taken into account by using positive values for motion in one direction and negative values for motion in the opposite direction. The diagram, **Fig. 9.4**, shows an example of two "vehicles" colliding and sticking together.

**The vehicles can be made to stick together by fitting one with a pin that sticks into some Plasticine on the other vehicle. A rebound collision occurs if the vehicles are fitted with repelling magnets.**

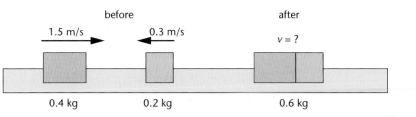

Fig. 9.4

If velocity to the right is taken to be positive then application of conservation of momentum gives:

$$0.4 \text{ kg} \times 1.5 \text{ m/s} + 0.2 \text{ kg} \times -0.3 \text{ m/s} = 0.6 \text{ kg} \times v$$

This gives a value for $v$ equal to +0.9 m/s, the positive sign showing that the velocity is to the right.

**When colliding objects stick together, the collision is always inelastic.**

In this type of collision the "vehicles" have less kinetic energy in total after they stick together than before they collide – some of it has been transferred to heat. When this occurs the collision is **inelastic**. In an **elastic collision** the total kinetic energy is conserved as well as the momentum.

Conservation of momentum also applies to **jet** and **rocket** propulsion. A jet uses oxygen from the surrounding air to burn its fuel, but a rocket carries its own oxygen supply, so the engines can be used outside the Earth's atmosphere.

If a balloon is blown up and released without its neck being tied, the air inside gains momentum as it is forced through the neck. To satisfy conservation of momentum the balloon must gain the same amount of momentum in the opposite direction, so that the total momentum remains zero. A similar thing happens in rocket propulsion, shown in the diagram, **Fig. 9.5**.

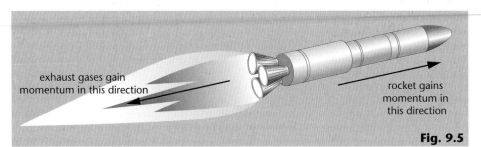

Fig. 9.5

**As a rocket burns fuel, its acceleration increases due to its decreasing mass.**

The exhaust gases gain momentum in the backwards direction as they are forced out of the rocket. The rocket gains the same amount of momentum in the forwards direction so that there is no overall change in momentum.

## Vehicle safety

The effect of a force in changing the momentum of an object depends on:

- the size of the force
- the time for which it acts.

> **KEY POINT**
>
> **force = change in momentum ÷ time**
> $$F = \Delta p / t$$

This statement is equivalent to **force = mass × acceleration**. In a road traffic accident a vehicle can be brought to a halt very rapidly. If the driver and passengers had the same deceleration they would experience large forces which could cause serious injury.

Injuries to the passengers in a vehicle are reduced by:

- **crumple zones** at the front and rear of the vehicle – these extend the time it takes for the vehicle to stop, reducing the deceleration of the vehicle
- **seat belts** – these restrain a passenger from colliding with parts of the vehicle, but stretch slightly to allow the passenger to carry on moving after the vehicle has stopped, reducing the deceleration of the passenger
- **air bags** – these inflate during a collision, reducing the pressure due to the retarding force acting on a passenger.

> When a vehicle stops, the passengers carry on moving until a retarding force acts on them.

---

**PROGRESS CHECK**

1. A bicycle travelling at 6 m/s increases its speed to 12 m/s in 12.0 s. Calculate its acceleration and the distance it travels while it is accelerating.
2. How does the horizontal motion of a projectile differ from its vertical motion?
3. Calculate the momentum of a 7 kg tenpin bowling ball travelling at a speed of 2.5 m/s.

1. 0.5 m/s² and 108 m;   2. The horizontal motion is at constant speed. The vertical motion is accelerated;   3. 17.5 kg m/s.

# 9.2  *Turning in a circle*

After studying this section you should be able to:

● *explain how the stability of an object is affected by the position of its centre of mass*
● *explain why an object in circular motion is accelerating*
● *calculate the size of the unbalanced force required to maintain circular motion.*

## Stability

All objects have a **centre of mass**, the point at which the force of weight is taken to act. For a solid object in the shape of a cuboid, the centre of mass is at the centre of the cuboid. For a human being it is in the region of the navel, and for a tyre it is in the space in the centre. The diagram, **Fig. 9.6**, shows the positions of the centres of mass of a traffic cone, a rubber ring and a person.

> In the case of a non-uniform object such as a hammer, the centre of mass is close to the heavy end.

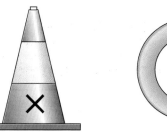 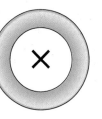 

Fig. 9.6

> The moment has a turning effect that returns the object to its rest position.

If an object is suspended freely, when it is at rest its centre of mass is directly below the point of suspension. In any other position the weight of the object has a **moment** (see section 2.3). The centre of mass of a shape in the form of a lamina can be found by:

● suspending the shape from a pin in a loose-fitting hole so that it is free to turn
● drawing a line vertically below the point of suspension when the object is at rest
● repeating this at other points
● the centre of mass is the single point where the lines cross.

This is shown in the diagram, **Fig. 9.7**.

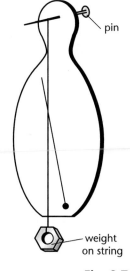

pin

weight on string

Fig. 9.7

The centre of mass of a bowling pin is higher than that of a traffic cone and it has a much narrower base. These factors combine to make it topple over when it is tipped. When a traffic cone is tipped it returns to its base.

The reason for this is shown in the diagram, **Fig. 9.8**.

> If the cone had a wider base, more tilt would be needed to make it topple.

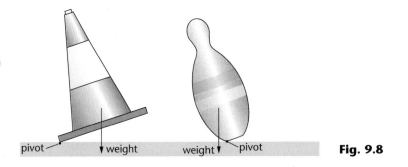

pivot ─ ▼weight    weight ▼ ─ pivot    **Fig. 9.8**

The traffic cone is **stable** and the bowling pin is **unstable** because:

- when the cone is tipped its weight exerts a restoring moment
- when the pin is tipped the line of action of its weight falls outside the base, so the moment of the weight makes the pin fall over.

## Going round in circles

AQA
Edexcel A    Edexcel B
NICCEA

Some artificial satellites occupy circular orbits and others are in elliptical ones. A satellite in a circular orbit maintains a constant speed, while that of a satellite in an elliptical orbit varies in a similar way to that of a comet.

> **KEY POINT**
>
> For a satellite in a circular orbit:
> $$\text{orbital speed} = \frac{2\pi \times \text{orbital radius}}{\text{time period}}$$
> $$v = 2\pi r/T$$

> An object moving in a circle has to keep moving towards the centre in order to stay the same distance away from it.

Motion in a circle at a constant speed is accelerated motion. This is because the direction of motion, and therefore the velocity, is constantly changing. This acceleration is called **centripetal acceleration** and is directed towards the centre of the circle.

> **KEY POINT**
>
> The relationship between centripetal acceleration, orbital speed and the radius of the orbit is:
> $$\text{acceleration} = \frac{(\text{orbital speed})^2}{\text{radius}}$$
> $$a = v^2/r$$

> To whirl a rubber bung around in a circle, you have to keep pulling on the string. If you let go of the string, the bung moves in a straight line at a tangent to the circle.

All acceleration requires an unbalanced force and in the case of circular motion the force is called the **centripetal force**. The unbalanced force on a satellite is gravitational and that on an orbiting electron is due to electrostatic attraction. The diagram, **Fig. 9.9**, shows these forces.

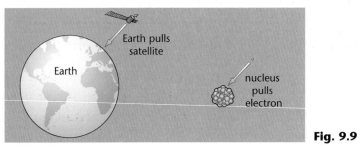

Earth pulls satellite

Earth

nucleus pulls electron

**Fig. 9.9**

It is important to remember that centripetal force is not a force that appears when an object moves in a circle. It is the unbalanced force that is acting on the object.

The size of the centripetal force on an object travelling in a circle:

- increases with increasing mass of the object
- increases with increasing speed of the object
- decreases with increasing radius of the circle.

**KEY POINT**

The relationship between the centripetal force and these variables is:

$$\text{force} = \frac{\text{mass} \times (\text{orbital speed})^2}{\text{radius}}$$

$$F = mv^2/r$$

**PROGRESS CHECK**

1. Explain why a half-full milk bottle is more stable than a full one or an empty one.
2. A person with a mass of 65 kg travels in a circle of radius 8.5 m on a fairground ride. Calculate the centripetal force and centripetal acceleration when the speed of the person is 12.0 m/s.
3. Suggest why cars sometimes leave the road when going round a bend in icy conditions.

3. The friction force is too small for the centripetal force required.
1. The half-full milk bottle has a lower centre of mass;    2. 16.9 m/s² and 1101 N;

# 9.3 Some effects of forces

**LEARNING SUMMARY**

*After studying this section you should be able to:*

- *describe how Hooke's Law applies to a spring*
- *explain the significance of the spring constant of a spring*
- *calculate the density of a material.*

## Using springs

OCR A $^B$
NICCEA
WJEC

For small stretching forces, springs and metal wires are **elastic** – they return to their original size and shape when the stretching force is removed. If a spring or metal wire is stretched beyond the **elastic limit** then the behaviour is **inelastic** and permanent deformation takes place.

The graph on p 134 shows the behaviour of a spring when subjected to an increasing stretching force. Metal wires show similar patterns of behaviour. Up to the limit of proportionality, springs and wires follow **Hooke's Law**.

Hooke's Law is very limited as it only applies to some materials over a limited range of forces. It cannot be used to make useful predictions.

**KEY POINT**

Hooke's Law states that:
the extension is proportional to the force provided that the limit of proportionality is not exceeded.

### Spring measurements

The **stiffness** or **spring constant** of a spring is a measure of how difficult it is to stretch it. The springs used in the suspension of a car and a washing machine are much stiffer than those used in retractable ballpoint pens.

**KEY POINT**

The spring constant of a spring, *k*, is defined by the equation:
force = spring constant × extension
$$F = k \times x$$

The diagram, **Fig. 9.10**, shows how to determine the spring constant as the gradient of a force–extension graph.

Springs **store energy** when they are stretched. When used in a car suspension, they absorb the energy of the vertical motion when the car goes over a bump in the road. The energy stored in a stretched spring is represented on a force-extension graph by the area between the graph line and the extension axis – this is the area shaded in the diagram above.

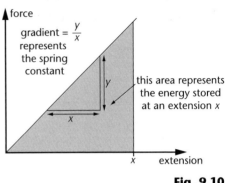

gradient = $\frac{y}{x}$ represents the spring constant

this area represents the energy stored at an extension *x*

**Fig. 9.10**

**KEY POINT**

If the extension of a spring or wire is linear:
energy stored = $\frac{1}{2}$ × force × extension
$$E = \frac{1}{2} \times F \times x$$

# Density

OCR A ᴮ
NICCEA

The density of a material is a measure of how close-packed the particles are. Measurements of both mass and volume are needed to calculate density.

**KEY POINT**

Density is calculated using the relationship:
density = $\frac{mass}{volume}$
$$p = m/V$$
Density is measured in g/cm³ or kg/m³.

Because the particles in a gas are more widespread than those in a solid or a liquid, gases have much smaller densities.

**PROGRESS CHECK**

1.  On a graph of force against extension for a metal wire, which part of the graph does Hooke's Law apply to?
2.  A 15 N weight stretches a spring by 0.22 m.
    Calculate:
    (a) the spring constant.
    (b) the energy stored in the stretched spring.
3.  Calculate the volume of iron in a 1 kg mass. The density of iron is 7.8 g/cm³.

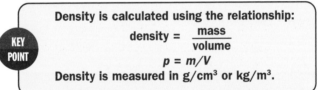

1. The straight line portion;   2(a). 68 N/m; (b). 1.65 J;   3. 128 cm³.

# Sample GCSE question

1. In a game of tennis, a ball leaves a racket travelling horizontally at a speed of 34 m/s from a height of 2.8 m above the ground. The diagram shows the path of the ball after it is hit.

34 m/s

2.8 m

The effect of air resistance on the ball can be neglected.

**(a)** What happens to the **horizontal velocity** and **vertical velocity** of the ball after it leaves the racket? **[2]**

*The horizontal velocity does not change ✓ and the vertical velocity increases ✓.*

> The fact that the horizontal velocity does not change is inferred from the statement that air resistance can be neglected.

**(b)** Explain the shape of the path followed by the ball. **[3]**

*The ball travels equal distances in equal time intervals horizontally ✓ but increasing distances vertically as its speed increases ✓ causing the path to become steeper ✓.*

**(c)** Calculate the time it takes for the ball to reach the ground.

The value of free-fall acceleration, $g$ = 10 m/s². **[3]**

$s = ut + \frac{1}{2}at^2$ *with $u = 0$* ✓
*so* $t = \sqrt{(2s/a)}$ ✓
$t = \sqrt{(2 \times 2.8\ m/10\ m/s^2)} = 0.75\ s$ ✓

> It is important to realise here that **vertically** the initial velocity is zero. A common error is to use data about the horizontal motion to answer a question about the vertical motion.

**(d)** How far does the ball travel horizontally in this time? **[3]**

*distance travelled = speed × time* ✓
$= 34\ m/s \times 0.75\ s$ ✓
$= 25\ m$ ✓

> The time is the only variable that has the same value for both the horizontal and the vertical motion.

**(e)** Calculate the **vertical velocity** of the ball as it reaches the ground. **[3]**

$v = u + at$ *with $u = 0$* ✓
$= 10\ m/s^2 \times 0.75\ s$ ✓
$= 7.5\ m/s$ ✓

> Remember, the correct unit is usually required to gain the final mark in a calculation.

# Exam practice questions

1. A Saturn V Moon rocket burns fuel at the rate of 13 000 kg each second.
   The exhaust gases have a speed of 2500 m/s.

   (a) Calculate the change in momentum of the exhaust gases in 1 s. [3]

   (b) The rocket has a mass of $2.7 \times 10^6$ kg. Calculate its change in speed each second. [3]

   (c) It continues to burn fuel at the same rate. Explain why its acceleration increases as more fuel is burned. [2]

2. The diagram shows two boys on roller blades facing each other.

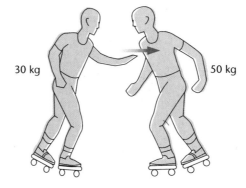

   30 kg        50 kg

   The smaller boy pushes the larger boy with a force of 150 N. The larger boy then moves off at a speed of 1.5 m/s.

   (a) What is the size of the force on the smaller boy? State its direction. [2]

   (b) Explain, in terms of momentum, why the smaller boy moves backwards. [3]

   (c) Calculate the initial speed of the smaller boy. [3]

3. A ball is travelling through the air.
   Its horizontal speed is 15 m/s.
   Its vertical speed is 10 m/s.

   Use a vector diagram to work out the speed of the ball through the air, and the direction of travel compared to a vertical line. [3]

4. A dart is thrown horizontally at a dartboard.

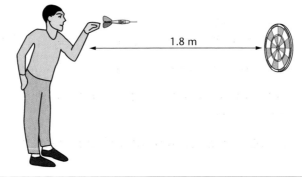

   1.8 m

# Exam practice questions

The horizontal speed of the dart is 12.4 m/s and the distance from the point of release to the dartboard is 1.8 m.

**(a)** Describe and explain the path of the dart as it travels through the air. **[3]**

**(b)** Calculate the time it takes for the dart to reach the dartboard. **[3]**

**(c)** Calculate the vertical distance travelled by the dart as it travels through the air.

The value of free-fall acceleration, $g = 10$ m/s$^2$. **[3]**

**5.** The diagram shows the circular orbits of two satellites.

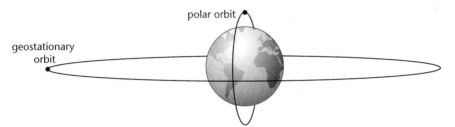

**(a)** Explain why an **unbalanced force** is required to keep a satellite in a circular orbit. **[2]**

**(b)** Describe the force that keeps a satellite in orbit. **[2]**

**(c)** What three factors determine the size of the unbalanced force required? **[3]**

**(d)** The orbital time of a satellite in a polar orbit is much less than that of a satellite in a geostationary orbit. Suggest two reasons why. **[2]**

**(e)** Here is some data about a surveillance satellite in a circular orbit.
radius of orbit $= 7.1 \times 10^6$ m
speed of satellite $= 7.5 \times 10^3$ m/s
mass of satellite $= 1.2 \times 10^3$ kg

    **(i)** Calculate the orbital time of the satellite. **[3]**

    **(ii)** Calculate the centripetal acceleration of the satellite. **[3]**

    **(iii)** Calculate the value of the unbalanced force that acts on the satellite. **[3]**

    **(iv)** What is the direction of this force? **[1]**

**6.** A golf ball has a mass of 0.046 kg.
After being struck by a golf club, it leaves the tee at a speed of 60 m/s.

**(a)** Calculate the change in momentum of the golf ball. **[3]**

**(b)** The golf club is in contact with the ball for a time of $4.0 \times 10^{-4}$ s.
Calculate the force exerted on the ball. **[3]**

**(c)** After hitting the ball, the club carries on moving with no noticeable change in speed. Suggest why. **[2]**

# Chapter 10 Particles

**The following topics are covered in this section:**

- **Atoms and nuclei**
- **Electron beams**
- **Particles in motion**

# 10.1 Atoms and nuclei

> **LEARNING SUMMARY**
>
> After studying this section you should be able to:
>
> - explain how the results of alpha particle scattering give evidence for the atomic model
> - explain whether an isotope is stable or unstable, by reference to the stability curve
> - describe the changes in the nucleus when an isotope undergoes beta decay.

## Farewell to the plum pudding

Edexcel A  Edexcel B
NICCEA
WJEC

Prior to the results of experiments carried out by Geiger and Marsden, who worked under the guidance of Lord Rutherford in 1911, the atom was thought to have the structure of a "**plum pudding**". The negatively-charged electrons were pictured as being evenly distributed within a positively-charged uniform mass.

> The alpha particles were detected by the flashes of light emitted when they hit a fluorescent screen.

Geiger and Marsden fired **alpha particles** from radon gas at thin gold foil. They then detected the alpha particles after they had been scattered. They found that:

- most of the alpha particles pass through the foil with no deflection
- some are deflected through a range of angles
- a small number are "back-scattered".

These outcomes are shown in the diagram, **Fig. 10.1**.

> The proportion of the alpha particles that are "back-scattered" is about 1 in 10 000.

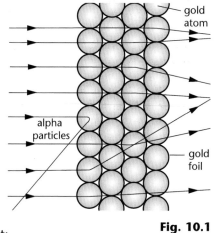

**Fig. 10.1**

Rutherford concluded from these results that:

- the atom is mainly empty space, allowing most of the alpha particles to pass through undeflected
- there are tiny regions of concentrated charge, which explains the large deflection of a small number of the alpha particles
- the charge on these regions must be the same sign as that on an alpha particle – positive – to cause the "**back-scattering**".

The currently accepted atomic model, shown in the diagram, **Fig. 10.2**, is that the atom has a central **nucleus** which contains almost all of the mass of the atom but occupies a very small fraction of the atomic volume. The nucleus has a positive charge which is balanced by the negative charge of orbiting electrons, so that overall the atom is neutral.

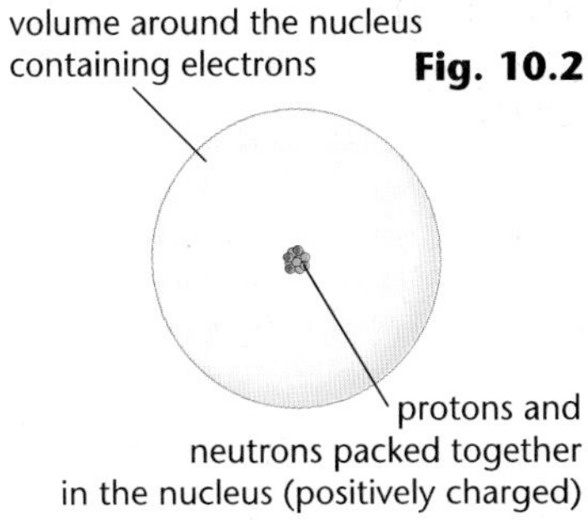

volume around the nucleus containing electrons **Fig. 10.2**

protons and neutrons packed together in the nucleus (positively charged)

> In scattering experiments the electrons do not affect the alpha particles, because of the vast difference in their masses.

Alpha particle-scattering experiments carried out using other materials show that the amount by which an alpha particle is deflected depends on:

- the closeness of its approach to the nucleus – the closer it gets, the bigger the force
- the charge on the nucleus – the greater the charge, the greater the force of repulsion
- the speed of the alpha particle – the faster it travels, the smaller the deflection.

## Nuclear stability

**Edexcel A** **Edexcel B**

Whether or not a nucleus is **stable** depends on the balance of protons and neutrons. The graph, **Fig. 10.3**, shows the relationship between the number of neutrons ($N$) and the number of protons ($Z$) for stable nuclei. Isotopes with nuclei in the regions of instability can decay in a number of ways according to their position on the graph:

regions of instability

number of neutrons ($N$)

number of protons ($Z$)

neutrons = protons

stability line

**Fig. 10.3**

- isotopes that are above the line have too many neutrons to be stable – they decay by $\beta^-$ **emission** when a neutron emits an electron, changing to a proton

> A positron is identical to an electron, but it has the opposite charge. A positron is very short-lived as it is annihilated when it meets an electron, resulting in two or three gamma ray photons.

- isotopes that are below the line have too many protons to be stable – they decay by $\beta^+$ **emission** when a proton emits a positron (an anti-electron), changing to a neutron
- isotopes with more than 82 protons in the nucleus usually decay by **alpha-emission**.

When a nucleus decays by emitting an alpha or beta particle the new nucleus that is formed often has too much energy to be stable. It loses this excess energy by emitting gamma radiation.

### How does beta decay affect the nucleus?

Electrons are **fundamental particles**, they are not made up of any other, smaller particles. This is not the case for protons and neutrons – each is made up of three smaller particles called **quarks**. There are a number of different types of quark, but only two are used to build protons and neutrons, the **up** quark and the **down** quark. The diagram, **Fig. 10.4**, shows the structure of a neutron and a proton and the charges on these quarks.

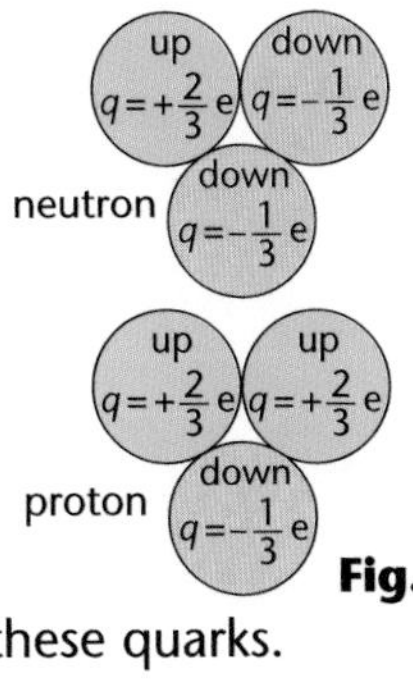

> Quarks cannot exist in isolation and they themselves may be made up of other particles.

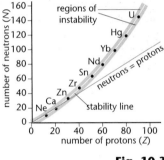

up $q = +\frac{2}{3}$ e  down $q = -\frac{1}{3}$ e

neutron  down $q = -\frac{1}{3}$ e

up $q = +\frac{2}{3}$ e  up $q = +\frac{2}{3}$ e

proton  down $q = -\frac{1}{3}$ e

**Fig. 10.4**

In β⁻ decay:

- a down quark in a neutron changes to an up quark, emitting an electron

- the atomic number (number of protons, $Z$) increases by one

- the mass number (total number of nucleons, $A$) is unchanged.

An example of β⁻ decay is the decay of strontium-90 into yttrium-90:

$$^{90}_{38}\text{Sr} \rightarrow ^{90}_{39}\text{Y} + ^{0}_{-1}\text{e}$$

**Check that these equations are balanced in terms of mass and charge.**

In β⁺ decay:

- an up quark in a proton changes to a down quark, emitting a positron

- the atomic number decreases by one

- the mass number is unchanged.

An example of β⁺ decay is the decay of carbon-11 to boron-11:

$$^{11}_{6}\text{C} \rightarrow ^{11}_{5}\text{B} + ^{0}_{+1}\text{e}$$

**PROGRESS CHECK**

1. How does "back-scattering" show that the atomic nucleus has the same sign as that on an alpha particle?
2. Explain why isotopes that lie above the stability line in fig 10.1.3 are sometimes described as being "neutron-rich".
3. Why is an electron called a fundamental particle?

1. In back-scattering the alpha particle is repelled by the nucleus. Objects with similar charges repel each other; 2. They have too many neutrons to be stable; 3. The electron is not made up of any other particles.

# 10.2  Electron beams

After studying this section you should be able to:

●  describe how an electron gun is used to produce a beam of electrons
●  calculate the kinetic energy of an accelerated electron
●  explain how an oscilloscope trace can be used to measure voltage and frequency.

## Producing a beam

The electrons in a metal wire are moving randomly at high speed. They are held within the confines of the wire by the attractive forces between them and the positive ions. Heating the wire gives the electrons more energy and they may gain enough to leave the wire in a process known as **thermionic emission**.

> The electrons need energy to do work against the attractive forces.

An **electron gun** uses thermionic emission to produce an **electron beam**. The diagram, **Fig. 10.5**, shows the structure of an electron gun.

In an electron gun:

● the low voltage supply can be either a.c. or d.c., its job is to heat the filament that forms the **cathode**

> The electrons are repelled from the negative cathode and attracted towards the positive anode.

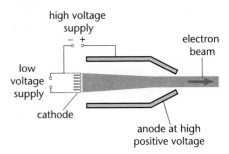

**Fig. 10.5**

● the high voltage between the **anode** and the cathode accelerates the electrons, pulling them away from the cathode, and focuses the beam
● the speed of the electrons in the beam depends on the voltage of the high voltage supply.

As the electrons leave the anode, their kinetic energy depends only on the voltage they have been accelerated through. The relationship between the kinetic energy of an electron and the accelerating voltage follows from the relationship between voltage and energy given in section 1.1.

> **KEY POINT**
>
> kinetic energy = electronic charge × accelerating voltage
> $$E_k = \tfrac{1}{2}mv^2 = e \times V$$

> The direction of the current is opposite to the direction in which the electrons are travelling.

A beam of electrons consists of moving charge and so it is an **electric current**. The size of the current at the anode can be calculated as the **rate of flow of charge**, $I = Q/t$.

> **KEY POINT**
>
> The current due to a beam of electrons is given by the relationship:
> current = number of electrons per second × electronic charge
> $$I = n \times e$$
> where $n$ is the number of electrons that pass a point in one second.

## Deflecting the beam

Edexcel A   Edexcel B
OCR A B

Electron beams are used:

- in televisions, oscilloscopes and computer monitors, where the beam causes fluorescence when it strikes a coated screen
- to produce X-rays – in an X-ray tube some of the energy from the beam is emitted as X-rays when the beam strikes a tungsten target.

A beam of electrons can be deflected by **magnetic fields** and **electric fields**. Televisions and computer monitors use magnetic fields, but in an oscilloscope the beam is deflected by an electric field between a pair of parallel plates. The diagram, **Fig. 10.6**, shows the deflection of a beam of electrons by an electric field.

> The motion of the beam between the plates is similar to that of a projectile. It follows a parabolic path when it is under the influence of the electric field.

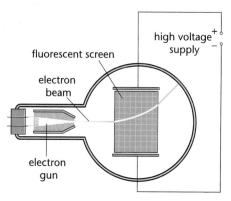

Fig. 10.6

Inkjet printers use a similar method for directing charged oil drops to the desired place on the paper.

In a television, a beam of electrons is directed at a screen that has three fluorescent coatings. These coatings emit light of the primary colours when the electrons strike them. The beam is made to move in a line across the screen in a process known as **scanning**. To produce a "picture", the electron beam scans the full screen with 625 lines and it completes 25 full-screen scans each second. The brightness of each colour at any point is controlled by changing the intensity of the part of the beam that strikes each of the three fluorescent layers. This is determined by the **modulation** of the signal.

## Using an oscilloscope

Edexcel A   Edexcel B
OCR A B

There are two sets of deflection plates in an oscilloscope:

- the X-plates deflect the beam horizontally
- the Y-plates deflect the beam vertically.

> When the time base is switched on, the dot appears as a line due to its rapid movement across the screen.

The X-plates are used to move the beam across the screen at a constant speed. The speed is set by the time base control. When the **time base** is switched off all that is seen on the screen is a dot. Switching on the time base enables the oscilloscope to plot a graph of **voltage** against **time**. The voltage is connected to the Y-plates and causes the electron beam to move up and down, according to whether the voltage is positive or negative. The **y-sensitivity** controls the amount of deflection for each volt of the input voltage.

The diagram, **Fig. 10.7**, shows an oscilloscope trace when an alternating voltage is connected to the input, with the time base switched on.

This shows a voltage that varies in value between +6 V and –6 V.

The frequency of an alternating voltage can be calculated by measuring the time for one oscillation. In this case one oscillation corresponds to a horizontal distance of 8 cm on the screen. The time period, $T = 8 \times 0.01 = 0.08$ s. It follows that the frequency, $f = 1/T = 12.5$ Hz.

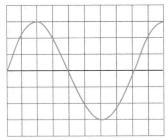

y-sensitivity = 2 V/cm    time-base = 0.01 s/cm

**Fig. 10.7**

---

**PROGRESS CHECK**

1. Explain how the current in an electron beam is affected by reducing the anode voltage.
2. Calculate the kinetic energy of an electron that has been accelerated through a voltage of 250 V. The electronic charge, $e = 1.6 \times 10^{-19}$ C.
3. Explain how you can tell whether a voltage is alternating or direct from an oscilloscope trace.

1. The current is reduced since fewer electrons pass through the anode each second.
2. $4.0 \times 10^{-17}$ J.    3. With an alternating voltage, the trace crosses the central line. With a direct voltage, it stays on one side of this line.

---

# 10.3 Particles in motion

**LEARNING SUMMARY**

After studying this section you should be able to:

- **explain the concept of absolute zero**
- **describe how changing the temperature of a gas affects its pressure**
- **calculate the energy transfer when an object is heated or cooled.**

## Gas pressure

Edexcel A    Edexcel B
OCR A $^B$

Gases consist of large numbers of particles in constant motion. The motion of an individual particle is **random**, it is constantly changing in both speed and direction as it collides with objects and other particles. Unlike the pressure due to a solid object, gases exert pressure equally in all directions.

**KEY POINT**
Gas pressure is a result of the forces exerted when the particles collide with the walls of the container.

> A common error at GCSE is to state that gas pressure is due to the collisions between the particles.

Experiments show that increasing the temperature of a gas results in an increase in the pressure. This is because:

- increasing the temperature increases the average speed of the particles
- this results in the collisions becoming more frequent
- the average force exerted during the collisions is greater.

When the pressure of a gas is plotted against its temperature, the result is a straight line, shown in **Fig. 10.8**. The straight line does not go through the origin because the Celsius temperature scale is not an **absolute** scale. On an absolute scale of measurement:

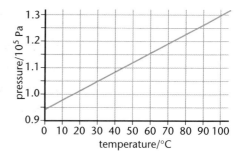

**Fig. 10.8**

The Celsius scale is a relative scale, temperatures are measured relative to the freezing point of water.

- zero is the minimum possible value
- 20 units has twice the value of 10 units.

This graph can however be used to locate the absolute zero of temperature and to set up a new scale that does measure temperature in absolute terms.

Real gases would liquefy before reaching this temperature. The absolute scale is based on an idealised gas model.

At the absolute zero of temperature the movement of the particles of a gas is at a minimum, and so is the pressure they exert. The diagram, **Fig. 10.9**, shows what happens when the graph is extrapolated to find the temperature at which a gas exerts minimum pressure.

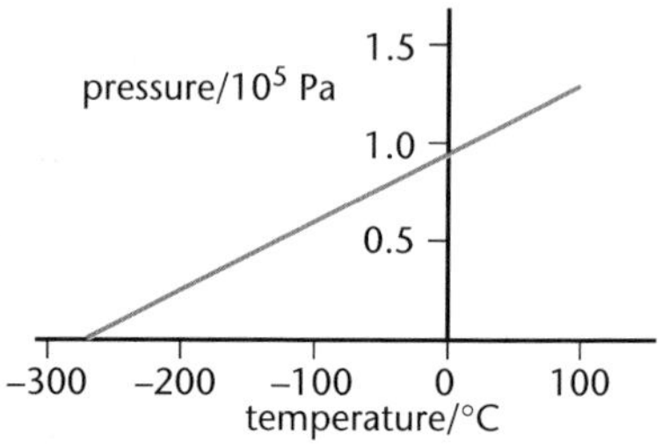

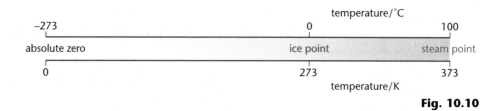

**Fig. 10.9**

The absolute zero of temperature turns out to have a value of –273°C. This is the minimum possible temperature – it is not possible to get any colder.

## Pressure and temperature

Edexcel A   Edexcel B
OCR A B

Knowledge of the absolute zero of temperature is used to establish the **kelvin** temperature scale. Celsius temperatures are converted to temperatures in kelvins simply by adding 273. The diagram, **Fig. 10.10**, shows the relationship between these temperature scales.

temperature/°C

| –273 | | 0 | | 100 |
|---|---|---|---|---|
| absolute zero | | ice point | | steam point |
| 0 | | 273 | | 373 |

temperature/K

**Fig. 10.10**

Doubling the temperature of a gas doubles the mean kinetic energy of the particles, but this does not mean that the speed is doubled.

When temperature is measured in kelvins:
- the mean kinetic energy of gas particles is proportional to their temperature
- the pressure of a gas is proportional to its temperature.

The last point means that when the kelvin temperature of a gas is doubled, its pressure also doubles.

**KEY POINT**

The relationship between the pressure of a gas and its kelvin temperature is:

$$\frac{P_1}{T_1} = \frac{P_2}{T_2}$$

This relationship can also be written as $P \times T = \text{constant}$.

## The gas equation

At a constant temperature, the pressure of a gas is inversely proportional to its volume (see section 2.3). The equation that describes this ($P_1V_1 = P_2V_2$) can be combined with the equation above to give a single equation. This equation applies to a situation where there are changes in the pressure, volume and temperature.

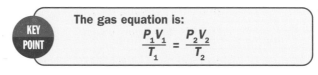

> **KEY POINT**
>
> The gas equation is:
> $$\frac{P_1V_1}{T_1} = \frac{P_2V_2}{T_2}$$

# Changing the temperature

**AQA**
**OCR A** [B]
**WJEC**

All objects have **internal energy** due to the movement of the particles. In solids and liquids there is a constant interchange between kinetic and potential energy as the particles vibrate and jostle against each other. In gases the internal energy is kinetic, and potential. The potential energy the particles in a gas have is gained when they change from a liquid to a gas.

> Temperature is a measure of the internal energy of the particles that make up an object.

Changing the temperature of an object involves changing its internal energy. This can be done by heating or cooling. The amount of energy depends on:

- the material the object is made from
- the mass of the object
- the temperature change.

> The term specific is used to refer to 1 kg of material.

These factors are all taken into account in the concept of **specific heat capacity**:

- specific heat capacity, $c$, is a property of a material
- it is the amount of energy required to change the temperature of 1 kg of the material by 1°C or 1 K.
- the units of specific heat capacity are J/kg °C or J/kg K.

> Although the values of temperatures differ by 273 when measured on the Celsius and kelvin scales, a temperature difference has the same value no matter which scale is used.

> **KEY POINT**
>
> The energy transfer when an object is heated or cooled is calculated using the relationship:
> **energy transfer = mass × specific heat capacity × temperature change**
> $$E = m \times c \times \Delta\theta$$
> where $\Delta\theta$ is the temperature change in °C or K.

**PROGRESS CHECK**

1. State two reasons why the pressure of a gas increases when its temperature rises.
2. At what temperature do the particles of a gas have twice as much kinetic energy as at 20°C?
3. What kind of energy do particles gain when they change from a liquid to a gas without a change in temperature?

1. The collisions are more frequent. The average force during collisions is greater because the average speed of the particles is greater.   2. 313°C or 586 K;   3. The particles gain potential energy.

# *Sample GCSE question*

**1. (a)** The diagram shows the "plum pudding" model of the atom.

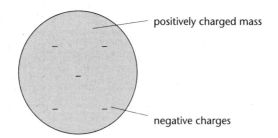

positively charged mass

negative charges

Explain how the position of the electrons on the currently accepted atomic model is different to that on the "plum pudding" model. **[2]**

> On the "plum pudding" model the electrons are positioned throughout the atom ✓ but on the currently accepted model they orbit the nucleus ✓.

*It is important here to use information given in the diagram. The diagram shows the electronic arrangement in the "plum pudding" model. The first mark is for describing the arrangement shown.*

**(b)** The next diagram shows some of the results when alpha particles are fired at thin gold foil.

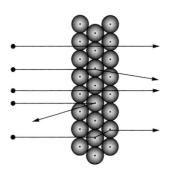

**(i)** Explain how this shows that the mass of the atom is not distributed uniformly. **[2]**

> Many alpha particles pass straight through, so they cannot have encountered any of the atomic mass ✓. Other particles are deviated by different amounts, so they cannot have passed through a uniformly distributed mass ✓.

*The cue word "explain" means that you need to give full reasons to support your answers. In this case the reasons are the different paths of the alpha particles that pass through the foil.*

**(ii)** Explain how this gives evidence about the charge on the atomic nucleus. **[2]**

> The charge on the nucleus must be the same as that on an alpha particle ✓ since some alpha particles are repelled ✓.

# Sample GCSE question

**(iii)** Suggest reasons why some of the particles that pass through the foil are deflected more than others. **[3]**

*Some pass closer to the nucleus than others ✓ so they experience a greater force ✓. The alpha particles could have different speeds, which would also affect the deflection ✓.*

**(iv)** Suggest how the results might differ if the experiment is carried out using a metal foil whose nuclei have less charge than those of gold. **[2]**

*There would be less deflection overall ✓ since the repulsive force would be smaller at the same distance ✓.*

The cue word "suggest" means that this may not be required knowledge, but the question requires you to apply your knowledge and understanding of other areas of physics to this particular situation.

# Exam practice questions

**1.** The graph shows the relationship between the number of protons and the number of neutrons in stable nuclei.

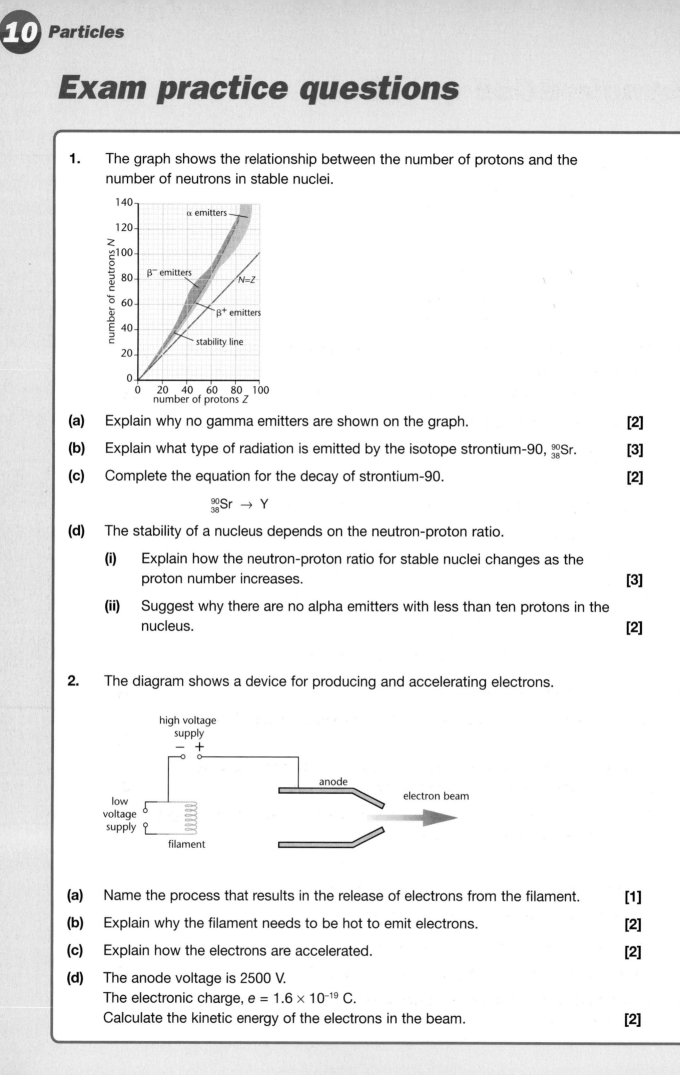

**(a)** Explain why no gamma emitters are shown on the graph. **[2]**

**(b)** Explain what type of radiation is emitted by the isotope strontium-90, $^{90}_{38}$Sr. **[3]**

**(c)** Complete the equation for the decay of strontium-90. **[2]**

$$^{90}_{38}\text{Sr} \rightarrow \text{Y}$$

**(d)** The stability of a nucleus depends on the neutron-proton ratio.

**(i)** Explain how the neutron-proton ratio for stable nuclei changes as the proton number increases. **[3]**

**(ii)** Suggest why there are no alpha emitters with less than ten protons in the nucleus. **[2]**

**2.** The diagram shows a device for producing and accelerating electrons.

**(a)** Name the process that results in the release of electrons from the filament. **[1]**

**(b)** Explain why the filament needs to be hot to emit electrons. **[2]**

**(c)** Explain how the electrons are accelerated. **[2]**

**(d)** The anode voltage is 2500 V.
The electronic charge, $e = 1.6 \times 10^{-19}$ C.
Calculate the kinetic energy of the electrons in the beam. **[2]**

# Exam practice questions

**3.** The diagram shows an oscilloscope trace.

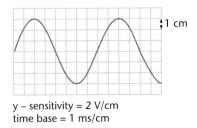

y – sensitivity = 2 V/cm
time base = 1 ms/cm

**(a)** How can you tell from the diagram that the input to the oscilloscope is an
alternating voltage? **[1]**

**(b)** **(i)** What is the maximum value of the input voltage? **[1]**

**(ii)** What is the minimum value of the input voltage? **[1]**

**(c)** Calculate the frequency of the input voltage. **[3]**

**4.** The diagram represents the particles of a gas in a cylinder.

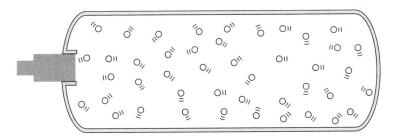

**(a)** Explain how the gas exerts a pressure. **[2]**

**(b)** Explain how the gas pressure changes when the temperature of the gas is
increased. **[3]**

**(c)** The pressure of the gas in the cylinder is $3.5 \times 10^5$ Pa when the temperature is
20°C. Calculate the pressure of the gas at a temperature of 50°C. **[3]**

**(d)** A label on the cylinder states that "this product must not be left in full sunlight".
Explain why this could be dangerous. **[3]**

**5.** A kettle is filled with 1.50 kg of water at a temperature of 10°C.

The kettle is switched on until the water boils at 100°C.

**(a)** Calculate the energy absorbed by the water.

The specific heat capacity of water = $4.2 \times 10^3$ J/kg °C. **[3]**

**(b)** Give three reasons why the kettle element needs to supply more energy than
the answer to **(a)**. **[3]**

# Exam practice answers

## Chapter 1 Electric circuits

1 (a) 1.6 V [1]
   (b) resistance = voltage ÷ current [1]
         = 1.6 V ÷ 0.5 A [1]
         = 3.2 Ω [1]
   (c) A filament lamp. [1]
      In a filament lamp the resistance increases as the filament gets hotter. [1]
   (d) The resistance increases. [1]
      The resistance at 6 V is 4 Ω. (or equivalent calculation) [1]

2 (a) (i)   Electron. [1]
      (ii)  Negative. [1]
   (b) Electrons move [1] from the ground [1] to neutralise the charge on the airframe. [1]
   (c) The aircraft can discharge when it lands [1] by charge passing [1] between the airframe and the ground [1]

3 (a) Current = charge ÷ time [1]
         = 300 C ÷ 60 s [1]
         = 5 A [1]
   (b) energy = charge × voltage [1]
         = 300 C × 12 V [1]
         = 3600 J [1]
   (c) power = current × voltage [1]
         = 5 A × 12 V [1]
         = 60 W [1]

4 (a) current = power ÷ voltage [1]
         = 8400 W ÷ 240 V [1]
         = 35 A [1]
   (b) The large current causes heating in the cables. [1] Thick cables have a low resistance so little heating occurs. [1]
   (c) It acts faster than a fuse [1]
      It is easily reset [1]
   (d) energy transfer = 8.4 kW × 3.5 h [1]
         = 29.4 kWh [1]
         cost = 29.4 × 7p = 206p [1]

5 (a) A [1]
   (b) B [1]
   (c) C [1]
   (d) C [1]
   (e) power = current × voltage [1]
         = 2.5 A × 12.0 V [1]
         = 30 W [1]

## Chapter 2 Force and motion

1 (a) (i)   B is the forwards push [1] of the wheel on the road. [1]
      (ii)  C [1]
      (iii) Wet leaves reduce the friction between the wheel and the road [1] which would cause the wheel to slip. [1]
   (b) (i)   Air resistance. [1]
      (ii)  60 N [1] forwards. [1]
      (iii) Acceleration = force ÷ mass [1]
            = 60 N ÷ 90 kg [1]
            = 0.67 m/s² [1]
      (iv)  The resistive force increases [1] The unbalanced force decreases [1] causing the acceleration to decrease. [1]

2 (a) (i)   Thinking distance is proportional to speed. [1]
      (ii)  As speed increases so does braking distance, but not in direct proportion. [1]
   (b) (i)   16 m [1]
      (ii)  21 m [1]
      (iii) 37 m [1]
   (c) (i)   Any two from: drugs, alcohol, tiredness, driver's concentration. [2]
      (ii)  Any two from: vehicle mass, condition of brakes, condition of road surface. [2]

3 (a) (i)   Acceleration = change in velocity ÷ time taken [1]
            = 30 m/s ÷ 3 s [1]
            = 10 m/s² [1]
      (ii)  Downwards. [1] The negative gradient of the graph shows that the acceleration is in the opposite direction to the upward velocity. [1]
      (iii) Force = mass × acceleration [1]
            = 0.020 kg × 10 m/s² [1]
            = 0.20 N [1]
   (b) (i)   1.5 s [1]
            This is when the sign of the velocity changes. [1]
      (ii)  Distance travelled = average speed × time. [1]
            = ½ × 15 m/s × 1.5 s [1]
            11.25 m [1]

4 (a) Pressure is due to collisions [1] between the gas particles and the container walls. [1]
   (b) The pressure increases [1] due to more frequent collisions. [1]
   (c) 4.5 × 10⁵ Pa × 0.015 m³ = 1.0 × 10⁵ Pa × V [1]
      V = 4.5 × 10⁵ Pa × 0.015 m³ ÷ 1.0 × 10⁵ Pa [1]
         = 0.0675 m³ [1]

## Chapter 3 Waves

1 (a) Any two from: infra-red, light, radio, ultraviolet, gamma. [2]
   (b) Ultraviolet. [1]
   (c) Infra-red. [1]
   (d) Sound/ultrasound. [1]
   (e) (i)   The device detects the echo. [1]
            The time is measured. [1]
            This is halved and multiplied by the speed of sound. [1]
      (ii)  Furniture would scatter the sound. [1] Giving multiple reflections. [1]

2 (a) Diffraction. [1]
   (b) The width of the gap [1] and the wavelength. [1]
   (c) The width of the doorway is approximately one wavelength for the sound wave [1] but is many wavelengths for light. [1]

3 (a) (i)   P [1] as they reached the detector first. [1]
      (ii)  There would be no S wave [1] as these do not travel through the Earth's core. [1]

(b) There is a shadow region directly opposite the centre of an earthquake. **[1]** S waves are not detected in this shadow. **[1]** Since transverse waves cannot travel in the body of a liquid **[1]** this shows that part of the core must be liquid. **[1]**

## Chapter 4 The Earth and beyond

1 (a) The satellite's orbit time is the same as the time it takes the Earth to rotate on its axis. **[1]** So the satellite remains above the same point on the Earth's surface. **[1]**

(b) Fixed aerials that point towards the satellite are used to receive the transmissions. **[1]** So the satellite needs to remain in a fixed position relative to the aerials. **[1]**

(c) (i) Arrow pointing towards the centre of the Earth. **[1]**

(ii) The size of the force increases. **[1]** The direction of the force changes so that it is always towards the centre of the Earth. **[1]**

(iii) When it is closest to the Earth. **[1]**

2 (a) Gravitational forces cause parts of the star to contract. **[1]** This causes heating. **[1]** A star is created when it is hot enough for fusion reactions to occur. **[1]**

(b) It formed from the remains of an exploding supernova. **[1]** Since these elements must have formed in a star after the main sequence. **[1]**

(c) (i) Hydrogen nuclei fuse together **[1]** to form the nuclei of helium. **[1]**

(ii) It will expand to become a red giant. **[1]** The outer layers will be flung off. **[1]** The core will then contract to become a white dwarf. **[1]**

3 (a) The wavelength of light received from other galaxies is lengthened **[1]** and shifted towards the red end of the spectrum. **[1]** This happens when objects are moving away from each other. **[1]**

(b) The wavelength of light received from Andromeda is shortened. **[1]** It shows "blue shift". **[1]**

(c) (i) Microwave radiation fills space. **[1]**

(ii) The Universe started with an enormous explosion. **[1]** Since then it has been expanding and cooling. **[1]** Stars and galaxies have formed from clouds of dust and gas. **[1]**

(d) Gravitational forces need to be big enough to stop the expansion **[1]** and cause the Universe to contract. **[1]** This will only happen if there is sufficient mass in the Universe. **[1]**

4 (a) (i) They cannot be seen with the naked eye. **[1]**

(ii) The movement of the moons **[1]** around Jupiter. **[1]**

(b) (i) The greater the orbital distance, the greater the period. **[1]**

(ii) The two inner moons are denser than the outer ones. **[1]**

(iii) They are larger. **[1]**

(iv) $6.2 \times 10^5$ s. **[1]**

## Chapter 5 Energy

1 (a) Electromagnetic induction. **[1]**

(b) (i) Moving the magnet. **[1]** Moving the coil. **[1]**

(ii) Any three from:
the number of turns on the coil, the area of the coil, the speed of movement, the strength of the magnetic field. **[3]**

(iii) Reverse the direction of movement. **[1]** Reverse the poles of the magnet. **[1]**

(c) (i) The current from the battery is d.c. **[1]** This creates a constant magnetic field. **[1]**

(ii) The current in the input coil is continually changing. **[1]** This produces a changing magnetic field in the output coil. **[1]**

(iii) The ratio of the voltages is 1:4. **[1]** The numbers of turns are in the same ratio, so there are 800 on the output coil. **[1]**

2 (a) Current in the conductors causes heating. **[1]** The water removes excess heat. **[1]**

(b) (i) They are much cheaper to install. **[1]** They are cheaper to run as the air acts as a coolant. **[1]**

(ii) There is not enough space for pylons. **[1]** Overhead conductors could be blown against buildings in windy conditions. **[1]**

(c) Any three from:
The power of the Sun's radiation is too low to make solar heating or solar cells economic.
There are few fast-flowing rivers and streams to generate hydroelectricity.
Rocks below the ground are not hot enough to generate steam to drive steam turbines.
There is not enough land where wind blows all the time to generate electricity in significant quantities. **[3]**

3 (a) A step-up transformer has more secondary turns than primary turns. **[1]** A step-down transformer has fewer secondary turns than primary turns. **[1]**

(b) Transformers are used to step up the voltage before electricity is transmitted, **[1]** and to step down the voltage before electricity is supplied to consumers. **[1]**

(c) The energy lost due to heating of the wires is proportional to the square of the current. **[1]** Distributing electricity at high voltages enables small currents to be used. **[1]**

(d) The voltage of a direct current cannot be changed. **[1]** Transformers allow alternating voltages to be stepped up and stepped down. **[1]**

4 (a) Any two from: Sun, wind, hydroelectric, tidal, geothermal, biomass. **[2]**

(b) Any two from: gas, coal, oil. **[2]**

(c) Gas-fired power stations are more efficient than coal-fired stations. **[1]**
So less carbon dioxide is released for the same amount of electricity production. **[1]**

(d) Coal-fired power stations preserve miners' jobs. **[1]**
If British coal is not burned then it takes away the livelihood of whole communities. **[1]**

(e) (i) Advantage: it does not use fossil fuels/does not cause atmospheric pollution. **[1]**
Disadvantages: any two from: unsightly, noisy, a large area of land is needed. **[2]**

(ii) There are few fast-flowing rivers and streams. **[1]** These are mainly in Wales and Scotland. **[1]** Obtaining energy from the tides is costly and unreliable. **[1]**

(iii) Advantages: it does not use up reserves of fossil fuel, **[1]** it does not release carbon dioxide into the atmosphere. **[1]**
Disadvantages: any two from: leaks can cause pollution of the atmosphere and local rivers and streams, it is very expensive to close down a nuclear power station safely, there is a constant problem of disposing of nuclear waste. **[2]**

**5** (a) Arrow drawn in the direction N to S. **[1]**
   (b) (i)   These sides are carrying a current at right angles to the direction of the magnetic field. **[1]**
      (ii)  These sides are carrying a current parallel to the direction of the magnetic field. **[1]**
   (c) (i)   Each force has a turning effect about the pivot (the central axis). **[1]** The moments both act in a clockwise direction. **[1]**
      (ii)  Reverse the current. **[1]** Reverse the direction of the magnetic field. **[1]**
      (iii) Increase the voltage/current. **[1]** Increase the strength of the magnetic field. **[1]**

## Chapter 6 Radioactivity

**1** (a) Beta particles are negatively charged. **[1]** They consist of high-speed electrons. **[1]** They are absorbed by a few mm of aluminium or other metal. **[1]**
   Gamma rays are not charged. **[1]** They are short-wavelength electromagnetic radiation. **[1]** Their intensity is reduced by thick lead or concrete. **[1]**
   (b) Here is a completed graph.

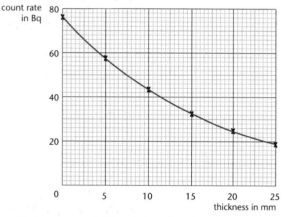

Marks are awarded for: choice of scale and correct labelling of axes **[1]** correct plotting of points **[2]** drawing of smooth curve. **[1]**
   (ii)  12.5 mm. **[1]**
         By using the graph to find the thickness required to halve the count rate from 76 Bq to 38 Bq. **[1]**
   (iii) 37.5 mm. **[1]**
         15 Bq is $1/8 = 1/2 \times 1/2 \times 1/2$ of 120 Bq. **[1]**
         So the thickness of lead needs to halve the intensity three times. **[1]**

**2** (a) Isotopes of an element have the same number of protons **[1]** but different numbers of neutrons. **[1]**
   Half-life is the average time it takes **[1]** for the number of undecayed nuclei to halve. **[1]**
   (b) (i)   So that the radiation penetrates the flesh **[1]** and can be detected by the camera. **[1]**
      (ii)  Alpha and beta radiation would be absorbed by flesh **[1]** and could cause damage to cells. **[1]**
   (c) 6 hours is long enough for the material to be active enough top be detected when it has circulated. **[1]** And short enough to minimise the risk of damaging the patient by exposure to radiation. **[1]**

**3** (a) (i)   New supplies of carbon-14 are constantly being absorbed **[1]** to replace that lost due to decay. **[1]**
      (ii)  As no new carbon-14 is being absorbed, the number of undecayed nuclei decreases. **[1]** So the number decaying each second decreases, as this is proportional to the number of undecayed nuclei present. **[1]**
   (b) (i)   500 Bq. **[1]**
      (ii)  Time taken for rate of decay to halve = 5 600 years. **[1]** Indicated on graph. **[1]**
      (iii) If the sample were 0.5 kg the rate of decay would be 60 Bq. **[1]** This gives an age of 6 600 years. **[1]**

**4** (a) Any two from: sodium chloride is soluble in water, gamma radiation can be detected on the surface, the half-life is long enough for water to flow through the pipes and the leak to be detected. **[2]**
   (b) Radioactive sodium chloride is dissolved in the water at the input. **[1]** When water has flowed all the way along the pipes, a detector is used to find the leak. **[1]**
   (c) (i)   A Geiger-Müller tube. **[1]**
      (ii)  The reading would go up. **[1]** Since radioactive water accumulates around the leak. **[1]**
   (d) Three half-lives **[1]** = 45 hours. **[1]**

## Chapter 7 Electronics

**1** (a) Logic gate 1 is OR **[1]**. Logic gate 2 is AND **[1]**.
   (b) The completed table is:

| input A | input B | input C | output |
|---------|---------|---------|--------|
| 0 | 0 | 0 | 0 |
| 1 | 0 | 0 | 0 |
| 0 | 1 | 0 | 0 |
| 1 | 1 | 0 | 0 |
| 0 | 0 | 1 | 0 |
| 1 | 0 | 1 | 1 |
| 0 | 1 | 1 | 1 |
| 1 | 1 | 1 | 1 |

Award one mark for each two rows correct **[4]**.
   (c) C **[1]**.
   (d) The alarm is on **[1]** as the OR gate gives a 1 output when both inputs are 1 **[1]**.

**2** (a) The resistance of the thermistor increases **[1]**. The voltage across the thermistor goes up and that across the variable resistor goes down **[1]**. The heater is switched on when the input to the NOT gate is below the threshold for it to be high (a 0) **[1]**.
   (b) The temperature at which the heater is switched on can be adjusted **[1]**.
   (c) The voltage across the thermistor = 3.0 V **[1]**. Resistance of variable resistor = $2/3 \times 45 \ \Omega$ **[1]** = 30 $\Omega$ **[1]**.

**3** (a) (i)   The logic gate **[1]**.
      (ii)  The alarm **[1]**.
   (b) The completed tables are:

| door switch | |
|-------------|--------|
| **input conditions** | **output** |
| door open | 1 |
| door closed | 0 |

| temperature sensor | |
|--------------------|--------|
| **input conditions** | **output** |
| less than 10°C | 0 |
| more than 10°C | 1 |

One mark for each table completed correctly **[2]**.
   (c) (i)   When the door is closed **[1]** or the temperature is below 10°C **[1]**.
      (ii)  Place a NOT gate **[1]** between the NOR gate and the alarm OR between each input and the NOR gate **[1]**.

**4** (a) The resistance of the LDR increases **[1]**. The output voltage increases **[1]**.

(b) The voltage across the fixed resistor = 2.0 V **[1]**. The resistance of the fixed resistor = 2/3 × 2000 Ω **[1]** = 1333 Ω **[1]**.

(c) The fixed resistor has a greater share of the voltage **[1]** so the output voltage decreases **[1]**.

**5** (a) 3 Ω **[1]**.

(b) The completed table is

| Resistor | Current in A | Voltage in V |
|----------|--------------|--------------|
| 6 Ω | 2.0 | 4.0 |
| 9 Ω | 1.0 | 6.0 |
| 12 Ω | 0.75 | 6.0 |
| 18 Ω | 0.25 | 6.0 |

1 mark for each row completed correctly **[4]**.

## Chapter 8 Using Waves

**1** (a) Ray to centre of lens continued straight **[1]**. Ray parallel to principal axis drawn through principal focus **[1]**. Upright arrow drawn where these cross **[1]**.

(b) 13.3 cm **[1]**.

(c) Any two from: object is inverted, image is upright; image is further away from lens than object; image is larger than object **[2]**.

(d) Projector or enlarger **[1]**.

(e) So that the image is seen the correct way up on the screen **[1]**.

**2** (a) The frequency of vibration when an object is displaced and allowed to vibrate freely **[1]**.

(b) When an object is forced to vibrate at its natural frequency **[1]** large amplitude vibrations build up **[1]**.

(c) (i) As the length of the wire increases **[1]** the natural frequency decreases **[1]**.

(ii) 260 Hz **[1]**.

(iii) 1.6 m **[1]**.

(iv) $v = f \times \lambda$ **[1]** = 260 Hz × 1.6 m **[1]** = 416 m/s **[1]**.

**3** (a) Diffraction **[1]**.

(b) A patch of light **[1]** the beams do not overlap **[1]**.

(c) Bright and dark bands **[1]**. Bright bands occur due to constructive interference **[1]**. The dark bands are due to destructive interference **[1]**.

(d) The difference in path length between two waves arriving at the same place **[1]**.

(e) A whole number of wavelengths or an even number of half wavelengths **[1]**.

**4** (a) Digital **[1]**.

(b) Analogue **[1]**.

(c) (i) Magnetic **[1]**.

(ii) The magnetic domains are realigned **[1]** from random to an ordered arrangement **[1]**.

(iii) As the tape moves, there is a changing magnetic field in the core **[1]**. This causes an alternating voltage in the coil **[1]** by electromagnetic induction **[1]**.

**5** (a) There is less spreading of the beam **[1]** so less energy is lost **[1]**.

(b) Space waves **[1]** as the others cannot travel through the Earth's atmosphere **[1]**.

(c) (i) Any two from: it takes 24 hours to orbit the Earth; the satellite remains above the same point on the Earth's surface; the orbit is equatorial **[2]**.

(ii) The receiving aerial can be fixed **[1]**.

(d) It receives waves **[1]**, amplifies them **[1]** and retransmits them **[1]**.

(e) Digital signals can carry more information **[1]**. They can be restored to their original condition **[1]** giving better quality reception **[1]**.

(f) Short wavelength waves have a high frequency and can carry a lot of information **[1]**. Short wavelength waves can travel through the ionosphere **[1]**. There is less diffraction at the dishes with short wavelength waves **[1]**.

## Chapter 9 Forces and their effects

**1** (a) Change in momentum = mass × change in velocity **[1]** = 13 000 kg × 2 500 m/s **[1]** = 3.25 × 10⁷ kg m/s **[1]**.

(b) Change in speed = change in momentum ÷ mass **[1]** = 3.25 × 107 kg m/s ÷ 2.7 × 10⁶ kg **[1]** = 12.0 m/s **[1]**.

(c) The mass of the rocket decreases **[1]** so for the same increase in momentum each second there is a greater increase in speed **[1]**.

**2** (a) 150 N **[1]** from right to left **[1]**.

(b) The total momentum remains zero **[1]**. The larger boy gains momentum directed towards the right **[1]** so the smaller boy gains an equal amount of momentum in the opposite direction **[1]**.

(c) 30 kg × $v$ = 50 kg × 1.5 m/s **[1]**, $v$ = 50 kg × 1.5 m/s ÷ 30 kg **[1]** = 2.5 m/s **[1]**.

**3** The completed vector diagram is **[1]**:

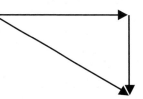

The speed of the ball is 18.0 m/s **[1]** at an angle of 56° to the vertical **[1]**.

**4** (a) The path is curved, with the slope of the curve increasing **[1]**. This is because the dart travels equal distances in equal time intervals horizontally **[1]** but increasing distances in successive equal time intervals vertically **[1]**.

(b) time = distance ÷ speed **[1]** = 1.8 m ÷ 12.4 m/s **[1]** = 0.145 s **[1]**

(c) $s = ut + \frac{1}{2}at^2$ **[1]** = ½ × 10 m/s² × (0.145 s)² **[1]** = 0.11 m **[1]**.

**5** (a) A satellite in a circular orbit is accelerating towards the Earth **[1]**. An unbalanced force is needed to cause this acceleration **[1]**.

(b) The gravitational pull **[1]** of the Earth on the satellite **[1]**.

(c) The mass of the satellite **[1]**, its speed **[1]** and radius of orbit **[1]**.

(d) The polar satellite has a greater speed **[1]** and a shorter orbital distance **[1]**.

(e)(i) $T = 2\pi r/v$ **[1]** = 2 × π × 7.1 × 10⁶ m ÷ 7.5 × 10³ m/s **[1]** = 5.9 × 10³ s **[1]**.

(ii) $a = v^2/r$ **[1]** = (7.5 × 10³ m/s)² ÷ 7.1 × 10⁶ m **[1]** = 7.9 m/s² **[1]**.

(iii) $F = m \times a$ **[1]** = 1.2 × 10³ kg × 7.9 m/s² **[1]** = 9.5 × 10³ N **[1]**.

(iv) Towards the centre of the Earth **[1]**.

**6** (a) Change in momentum = mass × change in speed **[1]** = 0.046 kg × 60 m/s **[1]** = 2.76 kg m/s **[1]**.

(b) Force = change in momentum ÷ time **[1]** = 2.76 kg m/s ÷

# Exam practice answers

$4 \times 10^{-4}$ s [1] $= 6.9 \times 10^{3}$ N [1].

(c) The mass of the club is large compared to that of the ball [1] so although its momentum changes by the same amount it hardly affects the speed [1].

## Chapter 10 Particles

1 (a) Gamma radiation is emitted by an excited nucleus [1] following the emission of alpha or beta radiation [1].

(b) Strontium-90 lies above the stability line [1] so it emits $\beta^-$ radiation [1]

(c) $^{90}_{38}$Sr $\rightarrow$ $^{90}_{39}$Y $+ ^{0}_{-1}$e 1 mark for each correct symbol on the right hand side of the equation [2]

(d) (i) For the least massive stable nuclei the neutron-proton ratio is 1 [1]. This increases as the nuclear mass increases [1] to a value of about 1.4 for the most massive nuclei [1].

(ii) For these nuclei to be stable the neutron-proton ratio is 1 [1], this would hardly change by emission of an alpha particle [1].

2 (a) Thermionic emission [1].

(b) The electrons need to gain energy to leave the metal [1], when the filament is heated the electrons gain energy [1].

(c) They are attracted to the positive anode [1] and repelled from the negative cathode [1].

(d) $E_k = e \times V = 1.6 \times 10^{-19}$ C $\times 2\,500$ V [1]
$= 4.0 \times 10^{-16}$ J [1].

3 (a) The trace shows both positive and negative voltages [1].

(b) (i) +6 V [1].

(ii) −6 V [1].

(c) Time for one cycle $= 8 \times 10^{-3}$ s. [1]. $f = 1/T = 1 \div 8 \times 10^{-3}$ s [1] $= 125$ Hz [1].

4 (a) The pressure is due to collisions [1] between the particles and the walls of the cylinder [1].

(b) The average speed of the particles increases [1] so the collisions are more frequent [1] and more force is exerted (on average) at each collision [1].

(c) $P_1/T_1 = P_2/T_2$ so $P_2 = P_1 \times T_2 \div T_1$ [1] $= 3.5 \times 10^5$ Pa $\times$ 323 K $\div$ 293 K [1] $= 3.9 \times 10^5$ Pa [1].

(d) The cylinder would absorb energy from the Sun's radiation [1] which would raise the temperature of the contents [1] and could cause the cylinder to explode due to the increased pressure [1].

5 (a) $E = m \times c \times \Delta\theta$ [1] $= 1.50$ kg $\times 4.2 \times 10^3$ J/kg °C $\times 90$°C [1] $= 5.67 \times 10^5$ J [1].

(b) Energy is needed to heat the element [1] and the casing [1] and some energy is lost to the surroundings [1].

# Formulae for relationships

This list shows the relationships that awarding bodies are not allowed to provide for candidates.

speed = $\dfrac{\text{distance}}{\text{time}}$

acceleration = $\dfrac{\text{change in velocity}}{\text{time taken}}$      $a = \dfrac{v-u}{t}$

force = mass × acceleration      $F = m \times a$

density = $\dfrac{\text{mass}}{\text{volume}}$

work done = force × distance moved in direction of force      $W = F \times x$

energy transferred = work done

kinetic energy = ½ × mass × speed²      $KE = \tfrac{1}{2} \times m \times v^2$

change in potential energy = mass × gravitational field strength × change in height      $GPE = m \times g \times h$

weight = mass × gravitational field strength      $W = m \times g$

pressure = $\dfrac{\text{force}}{\text{area}}$      $P = \dfrac{F}{A}$

moment = force × perpendicular distance to pivot

charge = current × time      $Q = I \times t$

voltage = current × resistance      $V = I \times R$

electrical power = voltage × current      $P = I \times V$

wave speed = frequency × wavelength      $v = f \times \lambda$

$\dfrac{\text{voltage across secondary}}{\text{voltage across primary}} = \dfrac{\text{number of turns on secondary}}{\text{number of turns on primary}}$      $\dfrac{V_p}{V_s} = \dfrac{n_p}{n_s}$

In addition, AQA require that candidates can recall the following relationships:

energy transferred = power × time      $E = P \times t$

cost of electricity = number of Units × cost per Unit

efficiency = $\dfrac{\text{useful energy transferred by device}}{\text{total energy supplied to device}}$

energy transferred = voltage × charge      $E = V \times Q$

Candidates for OCR specification A are required to recall the following relationships:

the equations of motion      $v = u + at$
$v^2 = u^2 + 2as$
$s = ut + \tfrac{1}{2}at^2$

momentum = mass × velocity      $p = m \times v$

energy transfer = mass × specific heat capacity × temperature change      $E = m \times c \times \Delta\theta$

efficiency = $\dfrac{\text{useful work or energy output}}{\text{total energy input}}$

| Centre number | |
|---|---|
| Candidate number | |
| Surname and initials | |

*Letts* **Examining Group**

**General Certificate of Secondary Education**

# Physics
# Higher Tier
# Paper 1

## Time: one and a half hours

### Instructions to candidates

Write your name, centre number and candidate number in the boxes at the top of this page.

Answer ALL questions in the spaces provided on the question paper.

Show all stages in any calculations and state the units.
You may use a calculator.

Include diagrams in your answers where this may be helpful.

### Information for candidates

The maximum mark for this paper is 90.

The number of marks available is given in brackets **[2]** at the end of each question or part question.

The marks allocated and the spaces provided for your answers are a good indication of the length of answer required.

 Where you see this icon you will be awarded marks for the quality of written communication in your answers.
This means, for example, that you should:
- write in sentences
- use correct spelling, punctuation and grammar
- use correct scientific terms.

| For Examiner's use only | |
|---|---|
| 1 | |
| 2 | |
| 3 | |
| 4 | |
| 5 | |
| 6 | |
| 7 | |
| 8 | |
| Total | |

**EDUCATIONAL**

**1 (a)** The diagram below shows a wind farm using several wind turbines to generate electricity.

**(i)** What change of energy takes place in a wind turbine?

........................................................................................... **[1]**

**(ii)** Name one disadvantage of using wind power.

........................................................................................... **[1]**

**(iii)** Is wind energy a renewable or non-renewable source?

........................................................................................... **[1]**

**(iv)** Each wind turbine is 74% efficient and generates 50 kW of electrical power.

Determine the input power to each wind turbine.

input power = ...................... kW **[3]**

**(b)** The diagram below shows part of a fairground ride.

carriage stops
here momentarily

ramp

H

14 m/s

600 kg

© Letts Educational 2005

*Letts*

The total mass of the carriage and its occupants is 600 kg.
The velocity of the carriage at the bottom of the ramp is 14 m/s.
The carriage climbs up a ramp and momentarily stops at a vertical height *H*.

**(i)** Calculate the kinetic energy of the carriage and its occupants at the bottom of the ramp.

kinetic energy = ............ unit: ......... **[4]**

**(ii)** The carriage comes to rest at the top of the ramp. What is the gain in gravitational potential energy of the carriage and its occupants? Assume there are no losses due to friction.

........................................................................................................ **[1]**

**(iii)** Calculate the height *H* of the carriage when it comes to rest. The gravitational field strength g = 10 N/kg.

*H* = ............ m **[3]**

**[Total: 14]**

**2**  The diagram below shows a 62 kg swimmer standing still on the edge of a diving board.

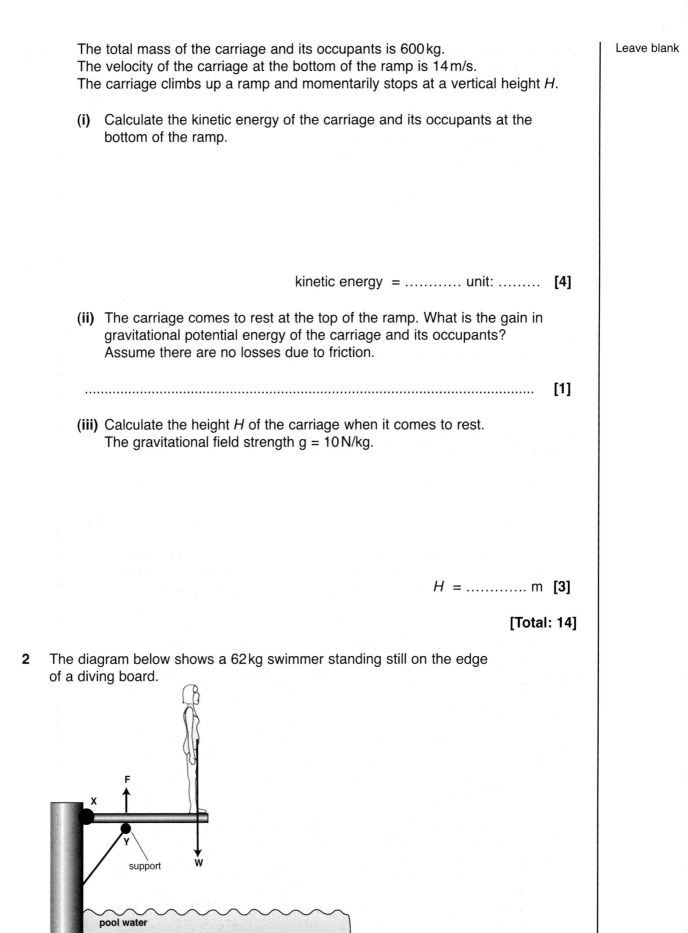

© Letts Educational 2005

**(a)** One of the forces acting on the swimmer is her weight. What other force acts on the swimmer as she stands still at the edge of the diving board?

.................................................................................................................... **[1]**

**(b)** Calculate the weight $W$ of the swimmer.
The gravitational field strength $g = 10\,\text{N/kg}$.

$W$ = ............. N  **[2]**

**(c) (i)** The weight $W$ of the swimmer exerts a moment about the point **X**.
State whether this moment is clockwise or anticlockwise.

.................................................................................................................... **[1]**

**(ii)** The diving board is resting on the support **Y**. Discuss whether the force $F$ at this support is greater than or less than the weight of the swimmer.

....................................................................................................................

.................................................................................................................... **[2]**

**(d)** The swimmer gently walks over the edge of the diving board and drops vertically down into the pool water below. The velocity against time graph below shows the motion of the swimmer in free fall and in the water.

© Letts Educational 2005

Use the graph:

**(i)** to describe the motion of the swimmer,
(One mark for correct spelling, punctuation and grammar.)

...........................................................................................................

...........................................................................................................

...........................................................................................................

............................................................................................ **[3+1]**

**(ii)** to calculate the initial acceleration of free fall of the swimmer.

acceleration = ............ m/s$^2$ **[3]**

**[Total: 14]**

**3** **(a)** Visible light is a transverse wave. Explain what is meant by a transverse wave.

.....................................................................................................

..................................................................................... **[1]**

**(b)** The diagram below shows part of a rear reflector of a car.

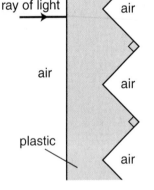

The reflector is made from red coloured plastic that is shaped as shown in the diagram so that the light from the rear is totally internally reflected.

**(i)** What happens to the speed of light when it travels from the air into the plastic?

................................................................................................................ **[1]**

**(ii)** Complete the path of the ray of light in the diagram. **[2]**

**(c)** The diagram below shows a piano being played in a room where the door has been left open.

Suggest why a person on the other side of the open door can hear the piano clearly even though he cannot directly see it.

................................................................................................................

................................................................................................................

................................................................................................................ **[3]**

**(d)** The diagram below shows two loudspeakers connected to the same a.c. supply.

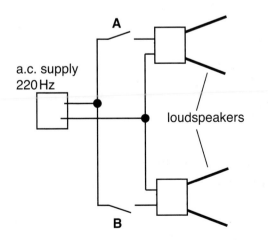

a.c. supply
220 Hz

loudspeakers

X

person listening
to sound from
loudspeakers

*Letts*

**(i)** With the switch **A** closed, and switch **B** open, a person at point **X** hears a very loud sound. The frequency of the sound is 220 Hz. Calculate the wavelength of the sound given the speed of sound in air is 340 m/s.

wavelength = ............. m  **[3]**

**(ii)** When both switches **A** and **B** are closed, the person suddenly hears no sound even though each loudspeaker is still emitting sound. Explain why this happens.

.....................................................................................................

.....................................................................................................

..................................................................................................... **[2]**

**[Total: 12]**

4  **(a)** Complete the ray diagram below to show what is meant by the focal length of a diverging (concave) lens. **[2]**

diverging lens

**(b)** The diagram below shows a converging (convex) lens used as a magnifying glass.

converging lens

F  O                          F        position of eye
                                       to see image

© Letts Educational 2005

The point marked **F** is the principal focus of the converging lens.
The object **O** lies between the principal focus and the centre of the lens.
The image produced by the lens is virtual.

Leave blank

**(i)** Explain what is meant by a virtual image.

................................................................................................

................................................................................................ **[1]**

**(ii)** Draw a ray diagram to locate the position of the virtual image formed by the lens. Label the image **I**. **[3]**

**(iii)** Apart from being a virtual image of the object **O**, state two other properties of this image.

................................................................................................

................................................................................................ **[2]**

**[Total: 8]**

**5** **(a)** The diagram below shows some sausages being heated under the heating element of a grill.

heating element

sausages

The heating element is at a temperature of about 1500 °C and the surface of the sausages is at about 220 °C. The electrical power rating of the heating element is 1.2 kW. The sausages take 10 minutes to cook.

**(i)** Name the process by which most of the heat reaches the sausages.

................................................................................................ **[1]**

**(ii)** Name the process by which the inside of the sausages become hot.

................................................................................................ **[1]**

© Letts Educational 2005

**(iii)** Calculate the cost of cooking the sausages.
Each Unit of electricity (kWh) costs 7.2 pence.

Leave blank

cost = ............. pence **[3]**

**(b)** The diagram shows two resistors connected to a 12V battery.

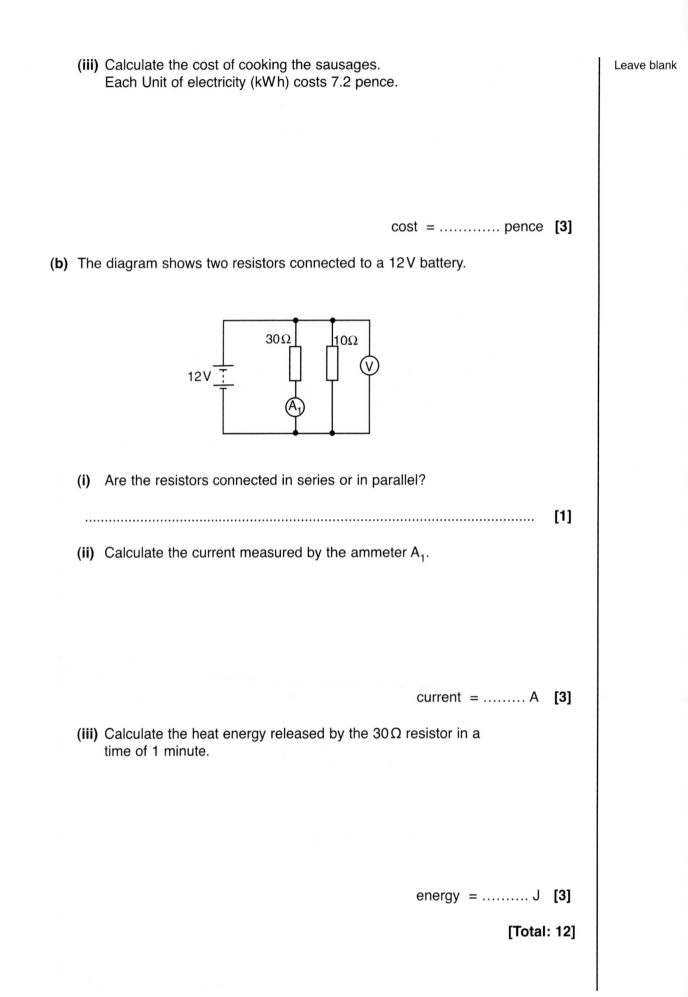

**(i)** Are the resistors connected in series or in parallel?

..................................................................................................... **[1]**

**(ii)** Calculate the current measured by the ammeter $A_1$.

current = ......... A **[3]**

**(iii)** Calculate the heat energy released by the 30Ω resistor in a time of 1 minute.

energy = ......... J **[3]**

**[Total: 12]**

**6** **(a)** The diagram below shows two coils made from insulated copper wires wrapped round a soft-iron rod.

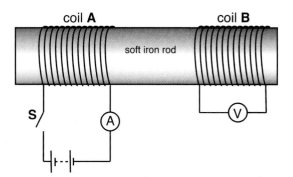

The coil **A** is connected in series with a switch **S**, an ammeter and a battery. The coil **B** is connected to a voltmeter.

The switch **S** is closed. The voltmeter connected to coil **B** shows a deflection and after a short period of time it shows no reading even though there is constant current shown by the ammeter. Explain these observations.
(One mark for a correctly ordered answer.)

........................................................................................................

........................................................................................................

........................................................................................................

........................................................................................................

........................................................................................................

........................................................................................................

........................................................................................................

........................................................................................... **[3+1]**

**(b)** A mobile phone charger unit has a transformer that steps-down the voltage from 230 V to 3.8 V. The primary coil has 5200 turns and an input current of 12 mA. The transformer is 100% efficient.

Calculate:

**(i)** the number of turns on the secondary coil,

turns = ............ **[3]**

(ii) the current in the secondary coil.

current = ............. A   **[3]**

**(c)** Electric power is distributed by the National Grid at a very high voltage of 400 000 V. Suggest why this voltage is so high.

.............................................................................................................

............................................................................................................. **[1]**

**[Total: 12]**

**7** Most of the carbon dioxide in the Earth's atmosphere contains atoms of the stable isotope carbon-12. A small percentage of the carbon dioxide also contains radioactive atoms of the isotope carbon-14.

**(a)** Explain what is meant by an isotope.

.............................................................................................................

............................................................................................................. **[2]**

**(b)** Explain what is meant by background radiation. Name one source for this radiation.

.............................................................................................................

............................................................................................................. **[2]**

**(c)** Carbon dioxide is absorbed by all living trees. The ratio of carbon-12 atoms to carbon-14 atoms in all living trees is a constant. When a tree dies it stops absorbing carbon-14 from the atmosphere. The carbon-14 already in the tree starts to decay. The half-life of carbon-14 atoms is 5600 years.

**(i)** An atom of carbon-14 may be represented as $^{14}_{6}$ C. In the nucleus of carbon-14, how many protons and neutrons are there?

protons: ............

neutrons: ............ **[2]**

**(ii)** An archaeologist discovers a wooden spear. A sample of 1.0 gram of carbon taken from the wooden spear gives an average of 22.5 counts per hour and a 1.0 gram sample of carbon taken from a living tree gives an average of 90 counts per hour. Assuming that the ratio of carbon-12 to carbon-14 atoms has remained constant since the spear was made, determine the age of the wooden spear in years.

age = ............. years **[3]**

**[Total: 9]**

**8** **(a)** Astronomers believe that the Sun and the planets were formed from dust particles and atoms, which were attracted together.

**(i)** State one major difference between the Sun and the planets in our solar system.

........................................................................................................... **[1]**

**(ii)** Name the force responsible for this attraction.

........................................................................................................... **[1]**

**(b)** Explain what is meant by an artificial satellite and suggest one of its uses.

...........................................................................................................

...........................................................................................................

........................................................................................................... **[2]**

*Letts*

**(c)** The diagram below shows an artificial satellite orbiting the Earth in a circular orbit of radius $6.7 \times 10^6$ m.

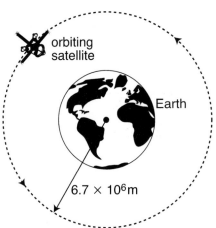

The speed of the satellite is constant at 7.7 km/s.

**(i)** Explain why the velocity of the satellite changes even though its speed remains constant.

.......................................................................................................

.................................................................................................... **[1]**

**(ii)** The centripetal acceleration $a$ of the satellite is related to its orbital speed $v$ and radius of orbit $r$ by the equation

$$a = \frac{v^2}{r}$$

**1** Explain what is meant by centripetal acceleration.

.................................................................................................... **[1]**

**2** Determine the centripetal acceleration $a$ of the satellite.

$a$ = ............. m/s$^2$ **[3]**

**[Total: 9]**

Letts

| Centre number | |
| --- | --- |
| Candidate number | |
| Surname and initials | |

*Letts* **Examining Group**

**General Certificate of Secondary Education**

# Physics
# Higher Tier
# Paper 2

## Time: one and a half hours

### Instructions to candidates

Write your name, centre number and candidate number in the boxes at the top of this page.

Answer ALL questions in the spaces provided on the question paper.

Show all stages in any calculations and state the units.
You may use a calculator.

Include diagrams in your answers where this may be helpful.

### Information for candidates

The maximum mark for this paper is 90.

The number of marks available is given in brackets **[2]** at the end of each question or part question.

The marks allocated and the spaces provided for your answers are a good indication of the length of answer required.

 Where you see this icon you will be awarded marks for the quality of written communication in your answers.
This means, for example, that you should:
- write in sentences
- use correct spelling, punctuation and grammar
- use correct scientific terms.

| For Examiner's use only | |
| --- | --- |
| 1 | |
| 2 | |
| 3 | |
| 4 | |
| 5 | |
| 6 | |
| 7 | |
| **Total** | |

*Letts*

**EDUCATIONAL**

**1** **(a)** The diagram below shows a child on a park swing.

**(i)** On the diagram above, state in which position **A**, **B**, **C** or **D** would the air resistance on the child be a maximum value.

..................................................................................................... **[1]**

**(ii)** Discuss the energy changes taking place as the child oscillates freely and comes to rest after some time.
(One mark for correct spelling, punctuation and grammar.)

.....................................................................................................

.....................................................................................................

.....................................................................................................

..................................................................................................... **[1+1]**

**(iii)** The rubber seat of the swing hangs from two sets of metal chains. Explain why the metal chains feel cold when the child grips them with both hands, but the rubber seat does not.

.....................................................................................................

.....................................................................................................

..................................................................................................... **[2]**

**(b)** The diagram below shows the main losses of thermal energy from a house.

Leave blank

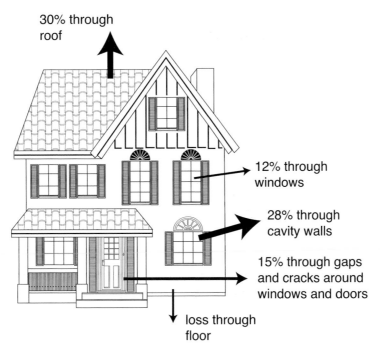

30% through roof

12% through windows

28% through cavity walls

15% through gaps and cracks around windows and doors

loss through floor

**(i)** Determine the percentage of heat loss from the floor of the house.

loss = ......................... % **[1]**

**(ii)** How can the heat loss through the flooring be reduced?

................................................................................................ **[1]**

**(iii)** The windows in the house are double-glazed. The diagram below shows a section through a double-glazed window.

trapped air

panes of glass

© Letts Educational 2005

*Letts*

Discuss how the heat loss through the double-glazed window is reduced compared to a window that has a single pane of glass.

......................................................................................................

......................................................................................................

...................................................................................... **[2]**

**(iv)** Explain why fitting shiny aluminium foil behind the radiators in the house would reduce heat loss.

......................................................................................................

...................................................................................... **[1]**

**[Total: 11]**

**2** **(a)** **(i)** Explain what is meant by the terms *thinking distance* and *braking distance* when describing the stopping distance of a car.

......................................................................................................

......................................................................................................

...................................................................................... **[2]**

**(ii)** Discuss one factor that affects the braking distance of a car.

......................................................................................................

......................................................................................................

...................................................................................... **[2]**

**(b)** A 940 kg car is travelling on a level road at a constant velocity of 20 m/s. The driver brakes suddenly and comes to rest in a time of 4.2 s after the brakes are applied.

**(i)** Calculate the deceleration of the car.

deceleration = ................ m/s² **[3]**

*Letts*

**(ii)** Calculate the braking force acting on the car.

Leave blank

force = ............................. N  **[3]**

The velocity against time graph for the braking car is shown below.

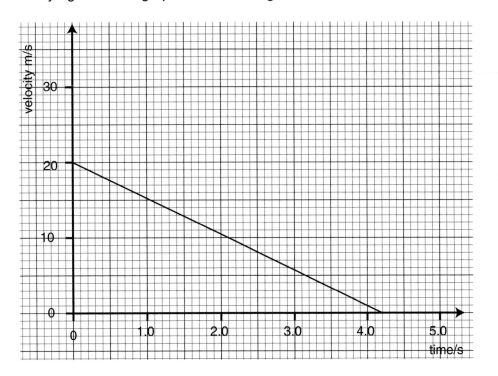

**(iii)** Use the graph to determine the braking distance of the car.

distance = ............................ m  **[3]**

**(iv)** Calculate the work done by the braking force.

work done  = ............... unit: .........  **[4]**

**(v)** Without any further calculations, suggest how the braking distance will change if the car was travelling at an initial speed of 40 m/s. You may assume that the braking force on the car remains constant.

.......................................................................................................................

.......................................................................................................................

....................................................................................................................... **[2]**

**[Total: 19]**

**3** **(a)** Suggest why sound waves cannot travel from the Earth to the moon.

.......................................................................................................................

....................................................................................................................... **[1]**

**(b)** Explain what is meant by ultrasound.

.......................................................................................................................

....................................................................................................................... **[1]**

**(c)** The diagram below shows an ultrasound scanner used in the aeronautics industry to detect cracks or flaws within the metal wings.

to oscilloscope   ultrasound scanner

A   B

2.8 cm

section through
the wing

crack within metal wing

The scanner consists of an ultrasound transmitter and a receiver. The signals are displayed on an oscilloscope.

**(i)** The transmitter is placed at position **A**. The oscilloscope trace below shows the transmitted pulse **T** and the reflected pulse **R** from the bottom surface of the metal sample.

voltage

T

R

time

Suggest why the reflected signal has smaller amplitude.

.................................................................................................

................................................................................................. **[1]**

**(ii)** The transmitter is now placed at position **B**. The oscilloscope trace below shows the transmitted pulse and the received pulses.

voltage

T

C

R

$1.5 \times 10^{-5}$ s

time

**1** Explain why there is an extra pulse **C**.

.................................................................................................

................................................................................................. **[1]**

**2** Use the information given by the oscilloscope trace to calculate the speed of the ultrasound in the metal.

speed = ......................... m/s **[3]**

**3** The frequency of the ultrasound is $3.6 \times 10^8$ Hz. Calculate the wavelength of the ultrasound.

wavelength = ........................ m   **[3]**

**[Total: 10]**

**4** **(a)** A rubber balloon acquires a positive charge when it is rubbed by a piece of cloth.

**(i)** In terms of electrons, explain how the balloon acquires a positive charge.

..................................................................................................

..................................................................................................   **[2]**

**(ii)** The diagram below shows two positively charged balloons which are placed alongside each other.

nylon string

On the diagram above, using arrows show the size and direction of the electric force experienced by each of the balloons.   **[2]**

**(b)** An electric kettle has 350 g of water at 20 °C. The kettle is switched on. The current in the heating element of an electric kettle operating at 230 V is 7.2 A.

Calculate:

**(i)** the resistance of the heating element of the heater,

resistance = ........................ Ω  **[3]**

**(ii)** the electrical power of the kettle,

power = ........................ W  **[3]**

**(iii)** how long it takes for the water to reach its boiling point (100 °C). It takes 4200 J of heat to change the temperature of 1 kg of water by 1 °C,

time = ........................ s  **[4]**

**(iv)** the charge flowing through the kettle.

charge = ................... unit: .......  **[4]**

**[Total: 18]**

**5** **(a)** Explain what is meant by the half-life of a radioactive substance.

...........................................................................................................

........................................................................................................... **[2]**

**(b)** The table below shows the half-lives of three emitters of β particles.

| β particle emitters | half-life |
|---|---|
| strontium-90 ( $^{90}_{38}$ Sr) | 28 years |
| lead-214 ( $^{214}_{82}$ Pb) | 27 minutes |
| carbon-14 ( $^{14}_{6}$ C) | 5700 years |

**(i)** Which part of the atom does the β particle come from?

........................................................................................................... **[1]**

**(ii)** State **two** properties of β particles.

...........................................................................................................

........................................................................................................... **[2]**

**(iii)** β particles are very dangerous to human beings. Explain why they are dangerous and state how you can minimise exposure from such a radiation.

...........................................................................................................

...........................................................................................................

........................................................................................................... **[2]**

**(iv)** Samples of the three β particle emitters have the same number of atoms. Which sample has the highest activity? Explain your answer.

........................................................................................................

........................................................................................................

........................................................................................................ **[2]**

**(v)** What fraction of the nuclei of lead-214 has **decayed** after a time of 81 minutes?

fraction = ................................. **[3]**

**[Total: 12]**

**6** **(a)** Explain why metals are good conductors of heat.
(One mark for a correctly ordered answer.)

........................................................................................................

........................................................................................................

........................................................................................................

........................................................................................................

........................................................................................................ **[3+1]**

**(b)** The gas atoms around us are constantly colliding with each other. The diagram below shows a helium atom making a head-on collision with a stationary carbon atom.

helium atom

carbon atom

500 m/s

$6.8 \times 10^{-27}$ kg

$2.0 \times 10^{-26}$ kg

**Before collision**

$v$

100 m/s

**After collision**

**(i)** Calculate the initial momentum of the helium atom.

momentum = ………….. kg m/s **[3]**

**(ii)** Use the information provided on the diagram above to determine the velocity of the helium atom after its collision with the carbon atom.

velocity = ………………………. m/s **[4]**

**[Total: 11]**

7　(a)　All logic gates use digital signals. With the aid of a sketch, explain what is meant by a digital signal.

voltage

time

.................................................................................................................

................................................................................................... **[2]**

(b)　The diagram below shows a logic gate.

A

B

output

(i)　Name the logic gate shown above.

................................................................................................... **[1]**

(ii)　Complete the truth table for the logic gate. **[1]**

| A | B | output |
|---|---|--------|
| 0 | 0 | |
| 0 | 1 | |
| 1 | 0 | |
| 1 | 1 | |

© Letts Educational 2005

**(c)** The diagram below shows a potential divider circuit consisting of a light-dependent resistor (LDR).

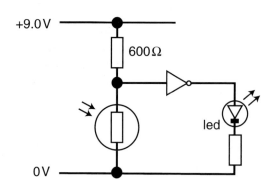

**(i)** State how the resistance of the LDR is affected by the intensity of light.

.................................................................................................................

................................................................................................... **[1]**

**(ii)** In sunlight, the resistance of the LDR is 1200 Ω. Calculate the voltage across the LDR.

voltage = ......................... V **[3]**

**(iii)** State whether or not the light-emitting diode (led) is lit.

................................................................................................... **[1]**

**[Total: 9]**

| Question | Answer | Mark |
|---|---|---|

**1 a i** Kinetic energy of the wind to electrical energy. **1**

   **ii** Generates energy only when it is windy or noise pollution. **1**

   **iii** Renewable. **1**

   **iv** efficiency = useful energy ÷ input energy **1**
0.74 = 50 ÷ input power **1**
input power = 50 ÷ 0.74 = 68 (kW) **1**

**Examiner's tip:**
The equation for efficiency is normally written as the ratio of energies, but remember power is energy per unit time. Hence the same equation applies to ratio of powers as well. In order to get the correct answer, it is sensible to set up the equation and then solve it in terms of the input power. There is no need to convert the kilowatts into watts because the final answer should be in kilowatts.

**b i** kinetic energy = $\frac{1}{2}$ × mass × velocity²

$(E = \frac{1}{2}mv^2)$ **1**

kinetic energy = $\frac{1}{2}$ × 600 × 14² **1**

kinetic energy = 58 800 ≈ 5.9 × 10⁴ **1**
unit: joule (J) **1**

**Examiner's tip:**
You should learn the units for all physical quantities. However, if you cannot remember the unit for kinetic energy, you can work it out from the equation: $E = \frac{1}{2}mv^2$. The unit will be: kg × (m/s)² → kg m²/s². This looks unfamiliar and complicated, but it is correct and the examiner will have to give you full credit. But it is easier to recall that the unit for energy is the joule.

   **ii** 58800 J **1**

**Examiner's tip:**
The energy is conserved. Since there are no losses due to friction (heat losses), all the kinetic energy is transferred into gravitational potential energy.

   **iii** gravitational potential energy = mass × gravitational field strength × height
(PE = $mgh$) **1**
58800 = 600 × 10 × H **1**

$H = \frac{58800}{6000} = 9.8$ m **1**

**Examiner's tip:**
As an A grade candidate, you must know how to rearrange an equation.

**2 a** There is a reaction from the board vertically upwards. **1**

**Examiner's tip:**
The swimmer is standing still; therefore the net force must be zero. You already know that the weight acts downwards hence the reaction must be equal but opposite to the weight.

**b** weight = mass × gravitational field strength
(W = mg) **1**
weight = 62 × 10 = 620 (N) **1**

**c i** Clockwise moment **1**

   **ii** The force F is greater than the weight **1**
because the distance of the force F from the pivot **X** is smaller. **1**

**Examiner's tip:**
moment = force × distance from pivot
The clockwise moment due to the weight is equal to the anticlockwise moment due to the force F. Since the distance of the force F from the pivot is smaller than the distance of the weight from the pivot, the force F must be larger than the weight.

**d i** The swimmer has a constant acceleration up to 1.5 s. **1+1**
After 1.5 s the swimmer is in the water and decelerates. **1**
*Any further detail.* **1**
(E.g.: The deceleration is not constant or the swimmer slows down because of resistance from the water)
**QWC mark:** The text is legible and the spelling, punctuation and grammar are accurate. **1**

**Examiner's tip:**
There are two marks for the first statement.
For writing down '*the swimmer accelerates*' you will only get one mark.

   **ii** acceleration = rate of change of velocity or acceleration = gradient of graph **1**

$a = \frac{10-0}{1.0}$ **1**

$a = 10$ (m/s²) **1**

**Examiner's tip:**
Always show your working. If you do the question by calculating the gradient from the velocity against time graph, then show the 'triangle used' on the actual graph. Examiners would expect working here since the acceleration for free fall is known by most candidates.

**3 a** The oscillations are at right angles to the wave direction.  **1**

**b i** The speed of light decreases as it enters the plastic.  **1**

**ii**

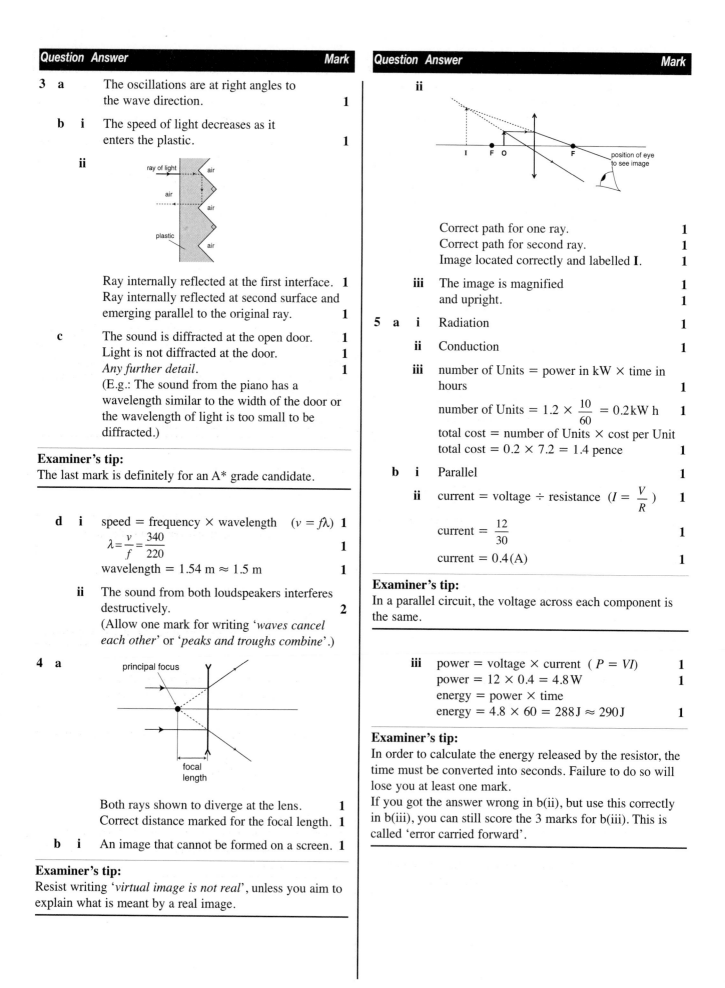

Ray internally reflected at the first interface.  **1**
Ray internally reflected at second surface and emerging parallel to the original ray.  **1**

**c** The sound is diffracted at the open door.  **1**
Light is not diffracted at the door.  **1**
*Any further detail.*  **1**
(E.g.: The sound from the piano has a wavelength similar to the width of the door or the wavelength of light is too small to be diffracted.)

**Examiner's tip:**
The last mark is definitely for an A* grade candidate.

**d i** speed = frequency × wavelength  ($v = f\lambda$)  **1**
$\lambda = \dfrac{v}{f} = \dfrac{340}{220}$  **1**
wavelength = 1.54 m ≈ 1.5 m  **1**

**ii** The sound from both loudspeakers interferes destructively.  **2**
(Allow one mark for writing *'waves cancel each other'* or *'peaks and troughs combine'*.)

**4 a**

Both rays shown to diverge at the lens.  **1**
Correct distance marked for the focal length.  **1**

**b i** An image that cannot be formed on a screen.  **1**

**Examiner's tip:**
Resist writing *'virtual image is not real'*, unless you aim to explain what is meant by a real image.

---

**ii**

Correct path for one ray.  **1**
Correct path for second ray.  **1**
Image located correctly and labelled **I**.  **1**

**iii** The image is magnified and upright.  **1**  **1**

**5 a i** Radiation  **1**

**ii** Conduction  **1**

**iii** number of Units = power in kW × time in hours  **1**
number of Units = $1.2 \times \dfrac{10}{60}$ = 0.2 kW h  **1**
total cost = number of Units × cost per Unit
total cost = 0.2 × 7.2 = 1.4 pence  **1**

**b i** Parallel  **1**

**ii** current = voltage ÷ resistance  ($I = \dfrac{V}{R}$)  **1**
current = $\dfrac{12}{30}$  **1**
current = 0.4 (A)  **1**

**Examiner's tip:**
In a parallel circuit, the voltage across each component is the same.

---

**iii** power = voltage × current  ($P = VI$)  **1**
power = 12 × 0.4 = 4.8 W  **1**
energy = power × time
energy = 4.8 × 60 = 288 J ≈ 290 J  **1**

**Examiner's tip:**
In order to calculate the energy released by the resistor, the time must be converted into seconds. Failure to do so will lose you at least one mark.
If you got the answer wrong in b(ii), but use this correctly in b(iii), you can still score the 3 marks for b(iii). This is called 'error carried forward'.

---

**6 a** The current in the coil **A** creates a magnetic field in the soft-iron rod. **1**

This magnetic field is changing when the switch is closed. **1**

This changing magnetic field links the coil **B** and hence a voltage is produced in coil **B**. **1**

Eventually, the constant current produces a constant magnetic field. There is no change in the magnetic field, hence there is no voltage produced in coil **B**. **1**

**QWC mark**: One mark awarded if points made in a logical order. **1**

**Examiner's tip:**
This is a tough question. It is testing whether or not you understand how a 'transformer' works.
Learn this section well.

**b i** $\dfrac{\text{voltage across primary}}{\text{voltage across secondary}} = \dfrac{\text{turns on primary}}{\text{turns on secondary}}$ **1**

$230 \div 3.8 = 5200 \div \text{turns on secondary}$ **1**

$\text{turns on secondary} = \dfrac{5200}{230} \times 3.8 = 86$ **1**

**Examiner's tip:**
Candidates often make a mess of rearranging this equation. You know that it is a step-down transformer. Hence the number of turns on the secondary will be smaller than that on the primary coil by a factor of

$\dfrac{230}{3.8} = 60.5$

Therefore, the number of turns on the secondary coil

must be $\dfrac{5200}{60.5} = 86$

**ii** input power = output power **1**
$230 \times 0.012 = 3.8 \times \text{current}$ **1**

$\text{current} = \dfrac{230 \times 0.012}{3.8} = 0.73\,(\text{A})$ **1**

**Examiner's tip:**
In a step-down transformer, the voltage is stepped down but the current increases. The current increases by the same factor as the ratio of the number of turns.
Hence current in the secondary coil
$= 12\,\text{mA} \times 60.5 = 730\,\text{mA} = 0.73\,\text{A}$

**c** High voltage means smaller current therefore less energy is lost in the overhead cables as heat. **1**

**7 a** An atom of an element, the nucleus of which has the same number of protons **1** but different number of neutrons. **1**

**b** Background radiation is radiation produced by external sources. **1**
Rocks (granite) or outer space or nuclear fall out or medical sources. **1**

**c i** protons = 6 **1**
neutrons (= 14 – 6) = 8 **1**

**ii** $90 \div 22.5 = 4$ **1**
number of half-lives = 2 **1**
age $= 5600 \times 2 = 11\,200$ (years) **1**

**Examiner's tip:**
The question is definitely something to do with half-lives. You can start with 90 counts per hour and keep dividing the answer by 2 and stop when you reach 22.5 count per hour. The number of times you have to do this will give you the number of half-lives. Therefore:
$90 \div 2 = 45 \rightarrow 45 \div 2 = 22.5$
one half-life    another half-life

**8 a i** Sun emits light whereas a planet reflects light. **1**

**ii** Gravitational force. **1**

**b** Man-made objects that orbit the Earth. **1**
Communications/weather monitoring/ spying. **1**

**c i** The direction changes, therefore the velocity changes. **1**

**Examiner's tip:**
Velocity has both size and direction. It is a vector quantity.

**ii1** Acceleration that is always directed towards a fixed point (in this case, the centre of the Earth). **1**

**ii2** radius = $6\,700\,000$ m and speed = $7700$ m/s **1**

$a = \dfrac{v^2}{r} = \dfrac{7700^2}{6700000}$ **1**

$a = 8.8\,(\text{m/s}^2)$ **1**

**Examiner's tip:**
Before substituting your numbers into the equation, you must first convert the radius into metres and the speed into metres per second.

| Question | Answer | Mark |
|---|---|---|

**1 a i**    Position **C**.      **1**

**Examiner's tip:**
The air resistance depends on the speed of the object. The drag force increases as the speed of the object increases. The question wants you to work out in which position the speed of the child is a maximum. The speed of the child is a maximum at position **C**.

   **ii**   *Any 2 from:*
Gravitational potential energy to kinetic energy from **A** to **C**.
Kinetic energy to gravitational potential energy from **C** to **D**.
Kinetic energy (or potential energy) to heat as the oscillations die out.    **1+1**
**QWC mark**: The mark is awarded for sentences without error in spelling, punctuation and grammar.    **1**

  **iii**    Metal is a good conductor of heat.    **1**
Heat flows from the hands to the chains, so the hands feel cold.    **1**

**b i**    15%    **1**

   **ii**    Loss through the flooring can be reduced by having a thicker carpet or an underlay.    **1**

  **iii**    Loss of heat by conduction is reduced    **1**
because air is a poor conductor of heat.    **1**

  **iv**    The shiny aluminium reflects radiation back into the house.    **1**

**2 a i**    The thinking distance is the distance travelled by the car whilst the driver is reacting.    **1**
The braking distance is the distance the car travels while the brakes are applied and the car stops.    **1**

**Examiner's tip:**
Candidates often forget to mention the important word **distance** in both definitions. It would be wrong to state that *'Thinking distance is how long it takes for the driver to react'* because the *'how long'* implies time. Always be very careful with definitions.

   **ii**    Any one factor from: road surface, efficiency of brakes, speed of car, mass of car and tyre conditions.    **1**
*Any further detail on one of the factors.*
(E.g: The greater the speed of the car, the greater is the braking distance or worn out tyres increase the braking distance of the car because of poor grip between the tyre and the road.)    **1**

**b i**    acceleration = rate of change of velocity    **1**
$a = \dfrac{0-20}{4.2}$    **1**
$a = -4.76 \text{ m/s}^2 \approx -4.8 \text{ (m/s}^2)$    **1**

**Examiner's tip:**
The velocity of the car decreases, therefore it is vital to have a **negative** sign for the deceleration. In order to avoid errors in examinations, write down the change in velocity as: 'final velocity – initial velocity'.

   **ii**    force = mass × acceleration    $(F = ma)$    **1**
$F = 940 \times -4.76$    **1**
$F = 4474\text{N} \approx 4500\,(\text{N})$ (Ignore the sign)    **1**

  **iii**    area under a velocity against time graph = distance    **1**
distance $= \dfrac{1}{2} \times 20 \times 4.2$    **1**
distance $= 42\,(\text{m})$    **1**

  **iv**    work done = force × distance moved in direction of force    **1**
work done $= 4474 \times 42$    **1**
work done $= 1.9 \times 10^5$    **1**
unit: joule (J) or Newton metre (Nm)    **1**

**Examiner's tip:**
It is always easier to write your answer in standard form, especially when the numbers are large (or small).

   **v**    The braking force is the same. Therefore the deceleration remains constant.
The time taken to stop is doubled.    **1**
braking distance $= \dfrac{1}{2} \times$ initial velocity × time
Since the time is doubled and the initial velocity is also doubled, the braking distance increases by a factor of four.    **1**

**Examiner's tip:**
This is a tough question. However, there are always clues left behind by examiners. This question is an extension of what you have already done in (iii). It is important to appreciate that the braking distance is directly proportional to the product of the initial velocity and the time taken for the car to stop. A common mistake would be to assume the braking distance is directly proportional to the initial velocity of the car and therefore end up with a wrong statement *'the distance doubles because the velocity doubles'*.

  

**3 a** There is a vacuum between the Earth
and the Moon. **1**

**Examiner's tip:**
It is very easy to use the wrong word and fail to secure a
mark. The key word in the marking scheme is '*vacuum*'.
You may use alternatives like: '*There is no air between the
Moon and the Earth*' or '*Sound needs air to travel*'.
Examiners tend not to like the use of the word '*space*' to
mean '*vacuum*'.

**b** A high frequency longitudinal wave (sound)
that humans cannot hear. **1**

**c i** Some of the wave energy lost. **1**

**ii 1** Ultrasound reflected from the crack. **1**

**ii 2** distance travelled by the ultrasound
$= 2 \times 2.8 = 5.6$cm **1**
speed = distance ÷ time **1**
speed $= \dfrac{0.056}{1.4 \times 10^{-5}} = 4000$ (m/s) **1**

**Examiners tip:**
A common error made with a question like this is
determining the total distance travelled by the reflected
wave. The wave is reflected by the crack. The total distance
travelled by the wave is twice the depth of the crack from
the surface of the metal. The answer for the speed is in m/s,
therefore do not forget to convert the distance into metres.

**ii 3** speed = frequency × wavelength ($v = f\lambda$) **1**
$\lambda = \dfrac{v}{f} = \dfrac{4000}{3.6 \times 10^{8}}$ **1**
wavelength $= 1.1 \times 10^{-5}$ (m) **1**

**Examiner's tip:**
To be an A grade candidate, you must be comfortable doing
calculations using standard form. Some candidates have
problems either recalling or rearranging the wave equation.
As an A grade candidate, you cannot afford to do this.

**4 a i** The rubbing action removes electrons
from the balloon. **1**
Electrons have a negative charge. The balloon
is left with a net positive charge. **1**

**ii**

Two arrows in opposite directions. **1**
Equal sized arrows (showing the same force
experienced by each balloon) **1**

**b i** resistance = voltage ÷ current ($R = \dfrac{V}{I}$) **1**
$R = \dfrac{230}{7.2}$ **1**
$R = 31.9\,(\Omega) \approx 32\,(\Omega)$ **1**

**ii** power = voltage × current ($P = VI$) **1**
$P = 230 \times 7.2$ **1**
$P = 1700$ (W) **1**

**iii** energy needed to change the temperature of
water by 1°C $= \dfrac{350}{1000} \times 4200$ **1**
energy needed to boil the water
$= \dfrac{350}{1000} \times 4200 \times (100 - 20)$ **1**
energy needed to boil the water
$= 1.18 \times 10^{5}$ J **1**
time = energy ÷ power $= \dfrac{1.18 \times 10^{5}}{1700}$ **1**
time $= 69$ (s)

**iv** charge = current × time ($Q = It$) **1**
charge $= 7.2 \times 69$ **1**
charge $= 500$ **1**
unit: coulombs or C **1**

**Examiner's tip:**
You can always work out what the unit for charge from the
equation $Q = It$. The unit for charge is:
ampere × second → As
If instead of the 'coulomb' you write down either 'ampere
second' or 'As', then the examiner has to give you full
credit for the correct physics.

**5 a** The half-life of a substance is the average time taken for half **1**
the number active nuclei to decay. **1**

**Examiner's tip:**
Quite often candidates will know that this definition involves something being halved. You have the choice of stating either '*number of active nuclei*' or the '*activity*'. Reference to the nuclei is important. It would be wrong to state that '*half-life is the time taken for the substance to halve its mass*'.

**b i** The nucleus. **1**

**Examiner's tip:**
Do not simply state that 'atoms' release β particles. It is very important to appreciate that radioactivity is to do with the changes taking place within the nuclei. Some candidates wrongly assume that the β particles are the orbiting electrons within the atoms.

**ii** *Any two from:* **1+1**
They are electrons.
Carry a negative charge.
Can be stopped by a thin sheet of aluminium.

**iii** They destroy living cells. **1**
Use a thin sheet of aluminium to shield from the source. **1**

**iv** The lead sample will have the largest activity. **1**
The nuclei decay in a short period of time (compared with the other samples). **1**

**v** 81 minutes = 3 half-lives **1**

fraction of nuclei left after 81 minutes =

$$\frac{1}{2} \times \frac{1}{2} \times \frac{1}{2} = \frac{1}{8}$$ **1**

fraction of nuclei that have decayed =

$$1 - \frac{1}{8} = \frac{7}{8}$$ **1**

**Examiner's tip:**
The question requires careful reading because it does not want you to just determine the fraction of nuclei **left** within the sample after 81 minutes. A mathematical way of getting the answer would be to use:

number of nuclei decayed = $1 - (\frac{1}{2})^n$,

where $n$ = number of half-lives.

**6 a** Metals are good conductors because of free electrons. **1**
These electrons can easily diffuse through the metal. **1**
Electrons gain thermal energy by colliding with the vibrating atoms at the hot end of the metal and transport this energy to the cooler atoms at the cold end. **1**
**QWC mark:** The mark is awarded for statements made in a logical order. **1**

**b i** momentum = mass × velocity **1**
momentum = $6.8 \times 10^{-27} \times 500$ **1**
momentum = $3.4 \times 10^{-24}$ (kg m/s) **1**

**Examiner's tip:**
Momentum is a vector quantity. It is therefore very important that you use 'velocity' in this equation and not 'speed'. Use of the term speed in the equation will lose you the first mark.

**ii** initial momentum = final momentum **1**
$3.4 \times 10^{-24} = (6.8 \times 10^{-27} \times v) +$
$(2.0 \times 10^{-26} \times 100)$ **1**
$v = (3.4 \times 10^{-24} - 2.0 \times 10^{-24}) \div$
$6.8 \times 10^{-27}$ **1**
$v = 206$ (m/s) **1**

**Examiner's tip:**
The key to doing this type of question is to set up an equation using the law of conservation of momentum and then solve for the unknown velocity $v$.

**7 a**

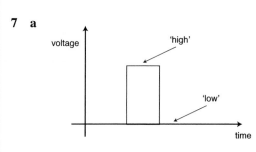

Correct sketch for a digital signal (see the sketch above). **1**
A digital signal has only two values or levels of voltage. **1**

**Examiner's tip:**
You can also state that a digital signal only has 'high' or 'low' values.

**b i** AND gate **1**

**ii**

| A | B | output |
|---|---|--------|
| 0 | 0 | 0 |
| 0 | 1 | 0 |
| 1 | 0 | 0 |
| 1 | 1 | 1 |

**1**

**c  i**

The resistance of the LDR decreases as the
intensity of light increases.  **1**

**ii**  total resistance $= 1200 + 600 = 1800\,\Omega$  **1**

current $= \dfrac{9.0}{1800} = 0.005\,\text{A}$  **1**

voltage $= IR = 0.005 \times 1200 = 6.0\,\text{V}$  **1**

**Examiner's tip:**
The current in a series circuit is the same. Use this idea to
calculate the voltage across the LDR and do not forget to
use the resistance of the LDR and not the resistor. Using
$600\,\Omega$ would give the wrong answer. If however, you show
all your working, then you would only lose the final mark.

**iii**  The led is not lit.  **1**

**Examiner's tip:**
The NOT gate inverts the input. The input voltage is 6.0 V,
which is closer to the 9.0 V than the 0 V. Therefore the input
to the gate is 'high'. The NOT gate changes this high input
into a low output. Consequently, the light-emitting diode is
not lit.

**HOW TO ASSESS YOUR GRADE**

The grid below suggests grades that you may expect to
achieve with different scores on these examination papers.
It is an indication only and does not imply that this is the
grade you will receive in the real examination.

| | |
|---|---|
| **A\*** | 148–180 |
| **A** | 120–147 |
| **B** | 96–119 |
| **C** | 70–95 |
| **D** | 48–69 |

# Index

numbers in italics refer to diagrams